Tracking the
CHILI LINE RAILROAD
to Santa Fe

MIKE BUTLER

Fonthill Media Inc.
www.fonthillmedia.com
office@fonthillmedia.com

First published 2020

Copyright © Mike Butler 2020

ISBN 978-1-63499-211-4

All rights reserved. No part of this publication may be reproduced, stored in a retrieval system or transmitted in any form or by any means, electronic, mechanical, photocopying, recording or otherwise, without prior permission in writing from Fonthill Media Inc.

Typeset in 10pt on 13pt Sabon
Printed and bound in England

ACKNOWLEDGMENTS

The majority of the historic photographs in this book come from the Richard L. Dorman Collection, which is now preserved by the Friends of the Cumbres and Toltec Scenic Railroad. My thanks go to Wes Pfarner, photo archivist for the Friends, for his assistance in providing me the photographs selected for this book. Thanks also to Cassie Mordini of the Aldo Leopold Foundation for the photograph of Aldo and Estella Leopold. Sources for the other historic photographs in the book are credited in the photo captions.

The Richard L. Dorman Collection is simply spectacular. There are 24,797 photographs in the collection from the 1880s to the present. Dorman was born in 1922 and was a World War Two B-24 bomber pilot, flying missions over the Pacific. After the war, he attended the University of Southern California and graduated in 1951 with a degree in architecture. He then began his architectural career in California, designing dozens of buildings. Dorman moved to Santa Fe in 1975 and continued his architectural career.

> After a ride on the Silverton train in 1973, Dorman became fascinated with the narrow gauge railroads of Colorado and New Mexico. He began collecting photographs, eventually amassing the largest extant narrow gauge photograph collection in the world, consisting of some 17,000 black-and-white images and over 7,000 color slides.[1]

During his years in Santa Fe, Dorman published several books on railroads with his own publishing firm, R.D. Publications, incorporating hundreds of photographs that he had collected. Dorman "also spent the better part of 30 years building a miniature railroad model that took up about 750 square feet of space in a specially built room adjacent to his home on Santa Fe's north-east side."[2]

Richard Dorman died on April 3, 2010, in Santa Fe after an eighty-seven-year life packed full of architecture and trains. A memorial service was held for him at First Baptist Church of Santa Fe, where he had been a deacon. This current book on the Chili Line could not have been possible without his photograph collection, and although having never met him, the author is deeply indebted to him.

While researching this book, the author consulted many sources (see Bibliography). However, any errors in this text are strictly his own.

Tracking the
CHILI LINE RAILROAD
to Santa Fe

CONTENTS

Acknowledgments   5
Introduction   9

  1. The Chili Line at Antonito   17
  2. Antonito to Tres Piedras   28
  3. Tres Piedras to Taos Junction   47
  4. Taos Junction to Embudo   61
  5. Embudo to Española   76
  6. Española to Buckman   89
  7. Buckman to Santa Fe   104

Epilogue   118
Appendix   121
Endnotes   122
Bibliography   127

INTRODUCTION

What a dream General William Jackson Palmer had—building a narrow-gauge railroad from Denver to Mexico City. Perhaps no one but Palmer could imagine such a thing back in 1870. It was a time of railroad barons dreaming of transcontinental railroads from the east to west coast. Yet Palmer wanted to build a north–south railroad. It was a revolutionary idea, which would eventually create the Chili Line to Santa Fe as part of Palmer's dream. Although the dream ended in Santa Fe, it was still a Herculean accomplishment to get the railroad built over the mountains from Denver to Antonito, Colorado, and then on into New Mexico and down to Santa Fe.

Palmer had the advantage of growing up around railroads. Born on September 18, 1836, in rural Delaware, he moved with his family five years later to Philadelphia where he began his schooling. In 1851, at the tender age of fifteen, he moved to western Pennsylvania to begin work with the Hempfield Railroad's engineering department. Thus began his career in railroading, which he was involved with the rest of his life. He was with Hempfield for four years, learning the operation of a railroad. In the summer of 1855, Palmer left for England for six months to study railroad operations there. He met with noted railroad engineers and visited railroads and coal mines. He then returned to Pennsylvania and went to work for the Pennsylvania Railroad. Palmer learned about running a railroad from John Edgar Thomson, the president of the "Pennsy." Palmer studied and wrote about converting railroad locomotives from wood-burning to coal-burning, since coal was a much more abundant fuel than wood.

The Civil War interfered with Palmer's railroad work in 1861. He joined the Union Army and had a distinguished service record, resulting in his eventual promotion to the rank of brevet brigadier general. In 1894, he was belatedly recognized for his service to the Union with a Congressional Medal of Honor.

After the Civil War, Palmer began work with the Kansas Pacific Railroad, which was building from Kansas City to Denver. He was treasurer of the construction company and was in charge of survey crews working across the Kansas plains. The Kansas Pacific reached Denver in the fall of 1870. Before that though, in 1869, Palmer struck out on his own, traveling south from Denver along the Rocky Mountain front, imagining the route that his dream railroad would take. He then boarded a train heading back east, looking for investors for his railroad. While on the train to Cincinnati, he met William Proctor Mellen, a New York attorney, and his daughter, Mary Lincoln Mellen. Palmer decided that he would marry Mary Lincoln Mellen, who was nicknamed "Queen." Palmer thought that was appropriate as he intended to be a railroad "king." Backed by William Mellen and other wealthy investors, Palmer incorporated his new Denver and Rio Grande Railroad on October 27, 1870, with a capital stock of $2,500,000.

The announced route for the new Denver and Rio Grande Railroad was "south from Denver to the Arkansas River near Pueblo, westward through the 'Big Canon of the Arkansas' across Poncha Pass into the San Luis Valley to the Rio Grande River, and thence along it to El Paso."[1] The estimated length of the route from Denver to El Paso was 875 miles. Palmer planned to build his railroad on a "narrow gauge" of 3 feet between rails, as opposed to the "standard gauge" of 4 feet 8.5 inches. He figured that this narrow gauge would allow his locomotives the ability to cover the steep grades and tortuous curves that would be encountered crossing the Rocky Mountains.

General Palmer began building his railroad right away. The first destination was the new town of Colorado Springs which Palmer founded on land that he had purchased. It was approximately 75 miles from Denver to Colorado Springs, and construction was completed in October 1871. On October 26, 1871, only a day shy of one year from the date of incorporation, the first ceremonial train left Denver for Colorado Springs with dignitaries on board. The 25,000-lb engine was named "Montezuma" for the ancient Aztec ruler of Mexico, and the two passenger cars were named "Denver" and "El Paso," representing the starting point and end point of the planned railroad, which hoped to eventually reach Mexico City.

At 15 miles per hour, it took the train five hours to reach Colorado Springs. After speeches and a banquet, the train headed back to Denver. Regular service between the two cities did not begin until January 1, 1872. Construction then began on the next leg of track to Pueblo, some 44 miles south of the Springs. Pueblo was reached on June 15, and construction continued west 35 miles along the Arkansas River to Labran (later named Florence) and the valuable coal fields there. Canon City, just 8 miles further west along the river, had to wait until July 6, 1874, for rails to reach their town due to disputes over costs with the railroad.

In his first annual report to stockholders in 1872, Palmer admitted that he was worried about the delays caused by building branch lines (such as the one to Labran). He stated that "It would be better to finish the whole line to El Paso, on the Mexican border, as it could rapidly be done in two to three years."[2]

By late 1872, the Kansas Pacific and the Atchison, Topeka and Santa Fe railroads were eyeing the same route along the Arkansas River, which the Denver and Rio Grande was pursuing:

> [In 1873] Palmer was very anxious to push on toward Santa Fe before any other road decided it was able to tap the New Mexican trade … he wrote that New Mexico already had a population of 110,000 which was perhaps twice that of Colorado when the road was begun. Just beyond reach lay the Santa Fe traffic, which … would nearly double the net earnings of the road if it could be reached.[3]

New Mexican businessmen were very enthusiastic about Palmer's plans. While they generally preferred to be linked to an east–west railroad (such as the Atchison, Topeka

Introduction

William Jackson Palmer, head of the Denver and Rio Grande Railway, *c.* 1870. (*Wikipedia, from Colorado Springs Museum*)

and Santa Fe line), it appeared that Palmer's Denver and Rio Grande would reach them first. In the fall of 1874, Palmer visited Santa Fe, "giving assurances that the Denver and Rio Grande would be completed to Trinidad (Colorado) by the following spring and New Mexico could expect a railroad shortly thereafter."[4]

Palmer's plans would soon face major challenges from his competitors though, and it would take another thirteen years before Santa Fe could be linked up with the Denver and Rio Grande by rail. Due to the competition along the Arkansas River, Palmer hatched an alternative plan to build south from Pueblo to Cucharas (Walsenburg) and west to La Veta, heading over Veta Pass to the San Luis Valley, where it would reach the Rio Grande and proceed south to New Mexico. Rails reached La Veta in 1876:

> The mountains were crossed via Veta Pass at 9,390 ft. altitude, which at the time was the highest point reached by any railroad in the United States, but only the first of several altitude records to be established in the development of Colorado's railroad system.[5]

Over the pass, Garland City (the town near Fort Garland) was reached on July 1, 1877. Construction continued west, reaching the new railroad town of Alamosa on the banks of the Rio Grande on June 26, 1878. Rails were then built south to the town of Antonito, reaching there in April 1880. Rails soon crossed the 5 miles to the Colorado–New Mexico border, and by the final day of 1880, they reached the New Mexico town of Española. There, however, things came to an abrupt halt, leaving Santa Fe waiting a short distance of 35 miles away. The Chili Line thus initially consisted of 90 miles from Antonito to Española. The remaining miles to Santa Fe would not be conquered until 1887.

The delay was a result of the war between the Atchison, Topeka and Santa Fe Railroad and the Denver and Rio Grande Railroad, each fighting to get their rail line built along the Arkansas River through the narrow Royal Gorge, which could only accommodate one set of tracks. The winning railroad would then have access north to the fabulously rich silver mines at Leadville. The Denver and Rio Grande was desperate for this route because it had lost the battle to secure the rights to Raton Pass, south of Trinidad, in 1878.

The AT&SF absolutely wanted Raton Pass for its transcontinental railroad. Its surveyors arrived at Raton Pass on the morning of February 29, 1878, and began staking and grading work. Half an hour later, the D&RG surveyors and crew arrived at the site. The competing crews faced each other and prepared for a fight. John A. McMurtrie, the chief engineer for the D&RG, decided to pull his men out and search for another route, an effort that proved unsuccessful. Losing the battle for Raton Pass meant that the D&RG became desperate to gain the route through the Royal Gorge.

Then began the "Royal Gorge War," with the two railroads fighting (literally) to gain the rights for trackage through the gorge. Both railroads had crews working at different points in the canyon. On the morning of April 20, 1878, the director of the D&RG crew had his men swim across the Arkansas River to the north bank ahead of the AT&SF crew, thus gaining the most favorable passage through the gorge. However, the AT&SF was then successful in gaining an injunction forbidding the D&RG to continue its work in the canyon. When the D&RG crew learned of this, they attacked the AT&SF crew, pushed them back down the canyon, and threw their tools in the river. Stymied, both railroads then continued their battle in the courts.

The court battle between the two railroads continued on for two years. Finally, in 1880, the two sides met in Boston, and came up with an agreement which came to be known as the "Treaty of Boston." The AT&SF agreed to give up the Royal Gorge route. In exchange, the D&RG agreed not to build a line east of Pueblo that would have competed with the AT&SF transcontinental line, and the D&RG agreed not to build any further south than Española, New Mexico, for a period of ten years. Thus, when the line to Española was completed on December 31, 1880, the Chili Line ended there. Service on the Chili Line was three days per week from Antonito to Española and did not expand to six days per week until the line was completed by the Texas, Santa Fe and Northern Railroad from Española to Santa Fe in 1887.

When knowledge of the Treaty of Boston became public, a group organized in Santa Fe to get a railroad line built from Santa Fe to Española. In December 1880, they organized the Texas, Santa Fe and Northern Railroad. In October 1881, the group sent a letter to investors, estimating plans and profits:

> Estimated annual revenues, based on 30 passengers each way and a daily movement of five cars inbound and two cars outbound of freight, would provide $132,000. After all expenses and interest, $30,000 was to be left for the stockholders.[6]

In 1882 the Texas, Santa Fe and Northern Railroad began work on the line with grading and bridge work. However, the railroad continually faced financial problems and had to stop work later in the year. The work was delayed on the line for nearly four years. Finally, work was resumed in August 1886. On October 21, the first rails were laid in

Introduction

East side of Santa Fe Plaza, 1866. It would be twenty-one more years until the Chili Line reached Santa Fe. (*National Archives #533174*)

Santa Fe, and the final spike was driven in Española late on Saturday night, January 8, 1887. The next day, Sunday, a train left Santa Fe for Española with 200 dignitaries on board. After a delay for a slight derailment, they finally reached Española "where dinner was served at the railroad hotel."[7]

With the Treaty of Boston still in effect, the Denver and Rio Grande Railroad was forced to transfer its passengers and freight to the Texas, Santa Fe and Northern Railroad at the Española depot. However, at least the "Chili Line" to Santa Fe was complete, and a passenger could board the train in Denver and arrive in Santa Fe approximately thirty-one hours later. Over seven hours of that trip was on the Chili Line. Covering the 125 miles from Antonito to Santa Fe took seven hours and forty-five minutes at an average speed of 16.2 miles per hour. The trip back from Santa Fe to Antonito took seven hours and fifteen minutes at an average speed of 17.3 miles per hour. This leisurely pace allowed passengers to fully enjoy the scenery along the way.

In 1889, ownership of the Texas, Santa Fe and Northern Railroad was transferred to the Santa Fe Southern Railway. In 1895, ownership was transferred to the Rio Grande and Santa Fe Railroad, which was a subsidiary of the Denver and Rio Grande. On July 1, 1896, the Denver and Rio Grande ran its own trains on the tracks to Santa Fe for the first time. Finally, on August 1, 1908, the Denver and Rio Grande formally purchased the Rio Grande and Santa Fe Railroad from its subsidiary.

Tracking the Chili Line Railroad to Santa Fe

The Chili Line had a rocky beginning. Unfortunately, things did not get a whole lot better over its lifetime until it was abandoned in 1941. However, its history is colorful, and it played a major role in opening the vast stretches of Northern New Mexico. The line received its nickname from one of the major crops grown along the line and shipped by the railroad—chili peppers (spelled "chili" or "chile," both are acceptable according to the dictionary, but "chili" became the preferred spelling of the railroad). The chili peppers were strung on cords ("ristras") and hung out in the sunshine to dry at the adobe homes along the line, and passengers on the Chili Line noted these colorful displays on their journey southward. Also, an old story relates that conductors on the railroad shouted out "chili!" to passengers when the train approached a meal stop.[8]

Chili ristras hang outside a home in San Ildefonso Pueblo along the Chili Line to Santa Fe. (*Bond House Museum, San Gabriel Historical Society, Española, NM*)

Map of the Chili Line. The Chili Line ran for 125.41 miles from Antonito, Colorado, to Santa Fe, New Mexico, reaching Española in 1880 and Santa Fe in 1887. (*Bond House Museum, San Gabriel Historical Society, Española, NM*)

1

THE CHILI LINE AT ANTONITO

El Rio Grande—the Great River—originates on Stony Pass, high in Colorado's San Juan Mountains just east of Silverton. It gushes down the steep mountain slopes east to the San Luis Valley, takes a turn south across the New Mexico border, plunges into the rift gorge of the Rio Grande del Norte National Monument, heads south down the center of New Mexico before turning east again along the Texas/Mexico border, and finally reaches its destination at the Gulf of Mexico, some 1,800 miles from its starting point. It is America's fifth longest river. It is literally the lifeblood of the arid San Luis Valley and New Mexico.

Colorado's San Luis Valley lies between the San Juan Mountains to the west, and the Sangre de Cristo Mountains to the east. At an average elevation of 7,500 feet above sea level, it stretches about 100 miles from north to south and 65 miles from east to west. Its high elevation limits the growing season for crops, but when the soil is irrigated from groundwater or by water from the Rio Grande, it produces abundant yields of potatoes, onions, cabbage, cauliflower, wheat, barley, peas, and hay. These are the products that would be shipped south to Santa Fe or north to Denver when the railroad was built down the valley.

The San Luis Valley was first seen by an American citizen when Lieutenant Zebulon Pike crossed over the Sangre de Cristo Mountains from the east in January 1807. Pike and his men were on an expedition authorized by President Thomas Jefferson to explore the Louisiana Territory, which had been purchased from France in 1803. Pike crossed a pass in the Sangre de Cristos and "looking westward down on the Great Sand Dunes, he wrote in his journal of seeing huge sand hills looking like a 'sea in a storm.'"[1]

Continuing on past the dunes, Pike reached a river that he thought might be the Red River but was actually the Rio Grande. Realizing that he was possibly now in Spanish territory, he built a defensive fort near the confluence of the Conejos River and the Rio Grande. "Pike's Stockade" was indeed soon surrounded by Spanish troops and Pike surrendered. He was taken to Mexico City, imprisoned, and interrogated, eventually being released back on American soil.

Mexico gained its independence from Spain in 1821, and in the 1830s, it began issuing land grants in the San Luis Valley, hoping to persuade Mexican families to settle there and discourage American settlement. However, Mexico was defeated in a war with the United States in 1848, and the San Luis Valley became part of America. San Luis became the first town settled in the valley in 1851 and is now recognized as the oldest town in Colorado.

With the San Luis Valley now in the United States, reaching the Rio Grande became General Palmer's goal when he created the Denver and Rio Grande Railroad. His goal was reached in June 1878, when his narrow-gauge tracks crossed the Rio Grande to the new town of Alamosa. Alamosa was created practically overnight when the town of Garland City was dismantled, packed up on the railroad, and shipped across the Rio Grande, where it was deposited and reconstructed.

From Alamosa, construction of the railroad southward was delayed for almost two years by the events of the "Royal Gorge War." Track laying began again in February 1880, reaching on March 30 what would become the town of Antonito, 28.9 miles south of Alamosa.

> [The tracks] followed the existing grade to Conejos, which the line bypassed by half a mile in order to set up the customary captive town on railroad-owned land, this one to be called Antonito, a mile to the south. Thus the railroad's owners were enriched by the sale of town lots while Conejos was bypassed and impoverished.[2]

Antonito ("little Anthony") took its name from the San Antonio River and San Antonio Mountain south of town. The town grew quickly, as it was at the junction of two D&RG branches—the line west to Chama, Durango and Silverton, and the line south to Santa Fe. Writing in 1891, A. R. Pelton noted the following of Antonito:

> [Antonito became] the principal shipping point for wool, hides, and pelts in this portion of the (San Luis) Valley ... more than half a million dollars in wool alone is shipped annually from this point. The markets of New Mexico are supplied with hay, oats and potatoes that are grown in this vicinity.[3]

Pelton also noted that the population of Antonito in 1891 was about 400 inhabitants; he observed:

> The Denver and Rio Grande Railroad has a fine stone depot built of black hard lava rock, and is said to be the most comfortable of any between this point and Pueblo. The company have also a round-house.[4]

San Luis Valley historian Virginia McConnell Simmons added to the inventory of known structures in Antonito:

> [It had] a section house, a bunk house, a sawmill, numerous saloons and gambling houses, a hotel, a newspaper, stores and three churches—Catholic, Presbyterian and Methodist vying for the degenerate souls of the local railroad men.[5]

One more structure in Antonito was an engine house. It served the Chili Line until a fire destroyed it in 1938. "Thus until the end of service in 1941, Chili Line motive power

The Chili Line at Antonito

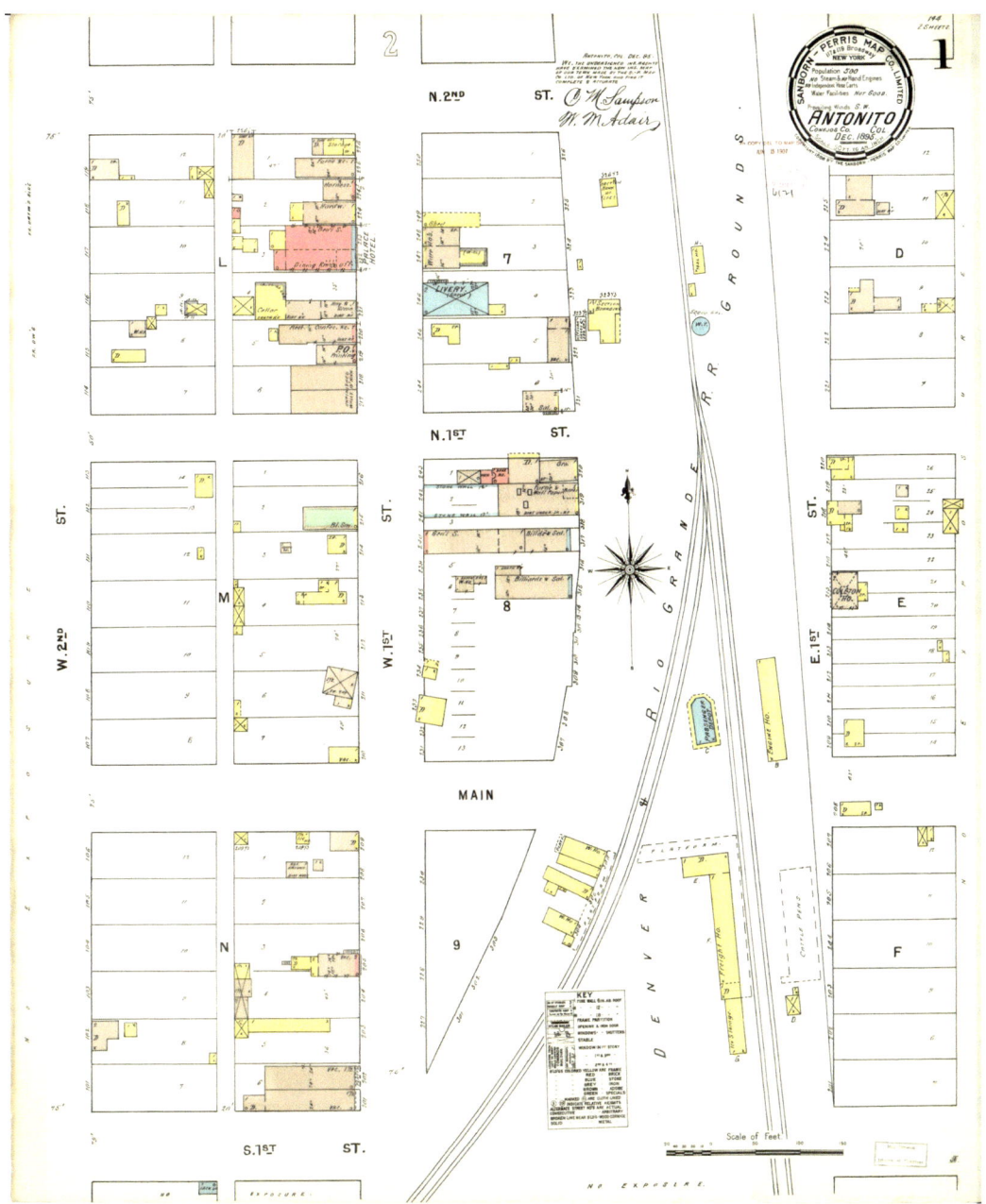

Sanborn map of Antonito, 1895. In the separation of the tracks, the tracks on the left led to Chama and Durango, and the tracks on the right led to Santa Fe. The passenger depot is shown in blue between the two sets of tracks. To the right of the depot is the engine house. South of the depot are freight warehouses and cattle pens. North of the depot, the blue circle is the 50,000-gallon water tank and to the left of that is a section boarding house and section bunk house. (*Library of Congress*)

This view is north toward the Antonito depot and water tank. It was taken on July 31, 1968, so the Chili Line was long gone, but the combined three rails for the standard and narrow gauges still remained. This set of tracks heads southwest to Chama, while the Chili Line tracks were behind the depot (now boarded up) to the right. (*Friends of the Cumbres and Toltec Scenic Railroad, Richard L. Dorman Collection, by N.P. Jenkins, RDS071-001*)

This view is south toward San Antonio Mountain with the Antonito depot on the right and the water tank on the left in this scene from June 29, 1965. (*Friends, Dorman Collection, RDS071-007*)

Robert L. Grandt took this photo at the Antonito water tank about 1920. This T-12 locomotive (no. 174) heads a consist of a water car, gondola, and three revenue cars. (*Friends, Dorman Collection, RD085-005*)

This undated view shows engine no. 473 taking water at the Antonito tank before heading south on the Chili Line. (*Friends, Dorman Collection, RD085-022*)

… started their journey to Santa Fe at Alamosa rather than Antonito each morning."[6] Antonito soon developed a rip-roaring reputation. Simmons notes:

> While railroad construction continued south and west, the workmen sought comfort in an entertainment capital of sorts in Antonito with its unusually high selection of bars and painted ladies lounging in the open doorways of their shacks.[7]

Antonito's population fluctuated wildly over the years with a high of 1,255 in 1950. The *WPA Guide To 1930s Colorado* noted that the population in 1940 (one year before the abandonment of the Chili Line) was 858, and "long sheds for the storage of potatoes and other vegetables adjoin the railroad station."[8] Today, the population of the town is about 800, and it serves as a departure point for the narrow-gauge Cumbres and Toltec Scenic Railroad, running west to Chama on the tracks of the former Denver and Rio Grande Railroad.

Engines 475 and 471 (behind) have arrived in Antonito after clearing snow from the Chili Line from Tres Piedras northward. This March 12, 1941, view was captured by photographer E. J. Foley, Jr. (*Friends, Dorman Collection, RDS077-099*)

Engine no. 473 prepares to head south from the Antonito depot to Santa Fe with a mixed train in this photo by Robert W. Richardson. (*Friends, Dorman Collection, RD085-019*)

On the left, engine no. 472 is waiting for engine no. 473 (behind), which has just arrived in Antonito from Alamosa. Railway Post Office (RPO) car no. 63 and coach no. 306 will be cut off for engine no. 473 to take to Santa Fe, and the remaining cars will be coupled behind engine no. 472 to take to Durango. This photo was taken by Robert W. Richardson on July 2, 1941. (*Friends, Dorman Collection, RD085-027*)

This photo at the Antonito water tank is undated, but it is probably late in the Chili Line era as the two engines (similar to the previous photograph) wait to take on water at the tank. (*Friends, Dorman Collection, RDS073-076*)

Since passengers at Antonito could either take a train to Durango or to Santa Fe, this sign was designed to direct passengers to the Chili Line. However, since "Taos" was misspelled as "Toas," it might have only added to their confusion. (*Friends, Dorman Collection, RDS071-228*)

The D&RG's sturdy lava rock depot still stands in Antonito in 2019, but it could use some loving care. There are rumors that restoration of the depot may occur in the future. (*Photograph by the author*)

The Cumbres and Toltec Scenic Railroad's depot and water tank are just south of the D&RG depot. Thousands of passengers ride this narrow-gauge train to Chama, New Mexico, each year from May through October. (*Photograph by the author*)

COLORFUL CHARACTERS OF THE CHILI LINE: ROBERT W. RICHARDSON

Occasionally as we travel along the Chili Line, we will meet characters whose stories add a dose of chili spice to the tale. The first of these is Robert W. Richardson, the founder of the Colorado Railroad Museum. Richardson very nearly missed his opportunity to ride the Chili Line. He was drafted by the U.S. Army in 1941, but his reporting date for duty was delayed for a year. During that time, Richardson decided to ride the rails of Colorado's narrow-gauge lines before they became extinct. His trip on the Chili Line began in Antonito on July 2, 1941, and after an overnight stay in Santa Fe, he rode the line back to Antonito on July 3. Only two months later, the Chili Line was abandoned forever. Though Richardson's trip was at the end of life for the Chili Line, his observations tell us quite a bit about what a ride on the line was like:

> Strangest of the Denver & Rio Grande Western's narrow-gauge lines was the 125-mile Santa Fe Branch, from Antonito, Colorado to the capital of New Mexico. It was like a train ride in a foreign country, earning it the nickname of the "Chili Line." This appellation was derived from the seasonal strings of drying hot peppers on the adobe houses of the few small communities along the branch ... At 8:30 a.m. the eight-hour southbound trip began. On the day that I rode on the train, there were no freight cars in the consist ... The open-platform coach had a bay window on each side, so the crew could keep an eye on the train ahead, somewhat like watching a train from a caboose cupola. Since this branch line seemed to be mostly curves, it was easy to watch the train from these bay windows.
>
> Passengers were few, including some railroaders, but mostly Hispanics. On this trip, mostly Spanish was to be heard, almost as if the train was in Mexico itself. Even the two restroom facilities were bilingually lettered "*Hombres*/Men" and "*Mujeres*/Women."[9]

As the train headed south, Richardson noted:

> This area was very dry, barren country, now and then relieved by a grove of stunted piñon trees ... It seemed that the track wound endlessly through dry, rough country made up of scattered volcanic rock. There were few trees, and at one point the line circled the rim of an ancient volcano, where there was a siding appropriately named "Volcano."[10]

Heading farther south, Richardson commented upon each of the main stops:

> It was 35 miles to the first town, Tres Piedras ("Three Rocks" in Spanish), with a station agent and a water tank as railroad amenities ... At Taos Junction, Milepost 336.52 (from Denver), there was a two-story depot, with upstairs living quarters for the agent, and again we exchanged mail. This time, the mail route was served by a rather battered automobile, which negotiated a dirt road to Taos, many miles to the east ... (At) Embudo, with a water tank and depot, was where we found the northbound train waiting for us. It had a few cars of freight in its consist ... After crossing the tributary Rio Chama ... we arrived at the lunch stop of the trip, in the

little town of Española … Beyond Española we passed through an Indian reservation, but there was no business there, just a couple of flag stops, sidings rusting away. And with 26 miles yet to go, we crossed the river (Rio Grande) and started a long climb … to the east. The railroad used long reverse loops through sandy areas, crossing dry washes, and at long last-suddenly-way below us was Santa Fe.[11]

Some of Richardson's observations may seem a bit "politically incorrect" for those of us in the twenty-first century, but Spanish is still widely spoken in Northern New Mexico, and bilingual signs can still be seen throughout the area. Sometimes, the lack of change can be a good thing.

2

ANTONITO TO TRES PIEDRAS

As soon as the tracklayers reached Antonito on March 30, 1880, they continued on south toward Tres Piedras, 34.7 miles away. "They laid second-hand 30-pound iron rail down to Tres Piedras which they reached on July 18."[1] This rail was cheaper than the 40-lb steel rail used to reach Antonito, and it reflected the cost-cutting measures that the D&RG used all along the Chili Line. Driving south from Antonito today on US Highway 285, travelers will essentially be following the route of the Chili Line to Tres Piedras.

Before continuing our journey south on the Chili Line, we will take a moment here to describe the types of steam locomotives used by the D&RG on the line.[2] This book is not meant to be a detailed survey of locomotives, freight cars, and passenger cars used on the Chili Line, but since captions on photographs used throughout the book often identify locomotive types, a brief primer is in order.

Early in the life of the Chili Line, the D&RG used T-12 class engines built by the Baldwin Locomotive Works between 1883 and 1884. These were ten-wheeled locomotives in a configuration of 4-6-0. This meant that there were four small non-powered pilot wheels (two on each side) at the front of the locomotive, followed by six larger power-driven wheels (three on each side) in the middle of the locomotive with no small trailing wheels at the back of the locomotive. These locomotives were numbered in the 160s and 170s.

This engine type and C-19 class 2-8-0 locomotives were used on the Chili Line until the early 1930s. These smaller locomotives were unable to pull freight cars unassisted up the 4-percent grade on Barranca Hill, so they had to be double-headed (two locomotives used). Therefore, the D&RG began to purchase larger locomotives which were able to handle the grade without a helper locomotive. The D&RG purchased fifteen of the K-27 series 2-8-2 locomotives from Baldwin, and they were numbered 450 to 464, but they saw limited use on the Chili Line and were mainly used on Marshall Pass in Colorado.

In 1923, the D&RG purchased ten K-28 class locomotives from the American Locomotive Company and numbered them 470 to 479. These 2-8-2 locomotives "weighed 68 tons and had a tractive effort of 27,540 pounds, which meant they could pull nearly any train up that 4% grade."[3] There are a number of photographs of these

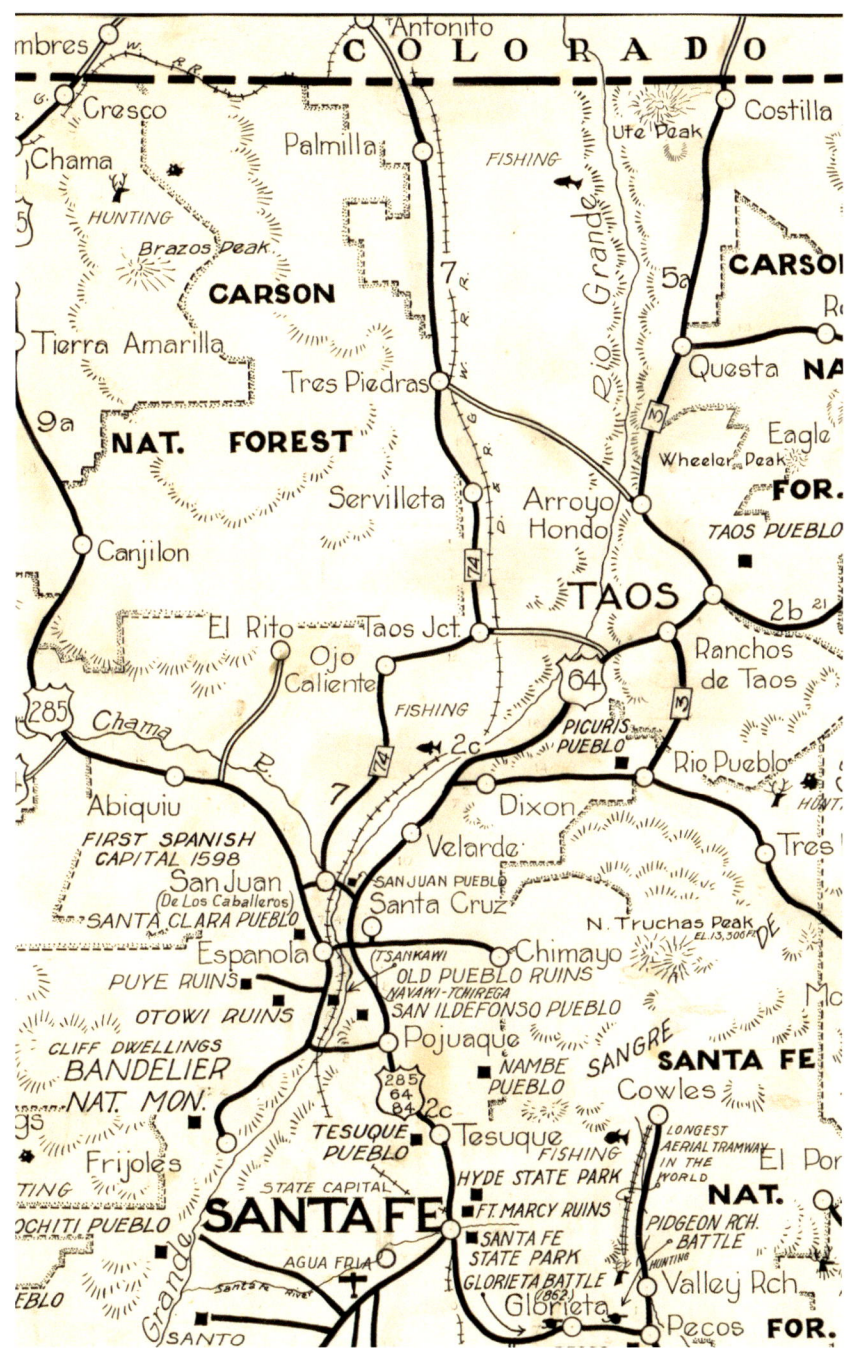

The 1939 Federal Writer's Project Map. This map (produced just two years before the Chili Line was abandoned in 1941) shows the general route of the railroad from Antonito to Santa Fe. Today, US Highway 285 runs from Antonito to Taos Junction, and New Mexico Highway 68 runs along the east side of the Rio Grande through Embudo, Velarde, and on to Española. In Española, US285 is once again picked up to head south to Santa Fe. (*Library of Congress*)

locomotives in this book. These locomotives were followed by a tender that carried 8 tons of coal and had a water tank capacity of 5,000 gallons.[4]

Passengers heading south on the Chili Line out of Antonito immediately noticed a large mountain looming in front of them. This is San Antonio Mountain, rising up to an elevation of 10,908 feet above sea level, and 3,000 feet above the surrounding plains. It is an extinct volcano, as are several other peaks in the vicinity. It encompasses about 9,000 acres. Due to its volcanic origin, the rock is very porous, so rain and snow seeps into the mountain rather than running down its slopes. This creates a hardship for wildlife searching for water on its forested slopes. Over the years, wildlife managers have set up water tanks around the mountain to take care of resident herds of antelope, deer and elk. New Mexico's native herd of elk had been exterminated by 1900 by hunters. Elk were then introduced from Wyoming in 1915, and it is from that herd that elk were introduced back to San Antonio Mountain in the 1930s. Today, the San Antonio elk herd is one of the largest in Northern New Mexico.[5]

Traveling south past San Antonio Mountain, the first stop reached by the Chili Line was Palmilla, 11.4 miles south of Antonito. An abundance of yucca plants here accounted for the name of the stop, as "*palma*" is the Spanish word for yucca. D&RG buildings at the site were located on the west side of the tracks and consisted of a section house, bunk house, coal house, and water tank.[6] The siding at Palmilla could accommodate up to thirty-two cars.[7]

The next stop south was Volcano, the highest elevation on the Chili Line at 8,487 feet above sea level. Volcano was obviously named after the extinct volcanic cones surrounding the area. It was 18.4 miles south of Antonito. There were no structures at Volcano, just a siding that could hold nineteen cars. Artist Birge Harrison rode the Chili Line in 1884 to his new home in Española. Writing in the May 1885 issue of *Harpers Monthly Magazine*, Harrison described the country between Palmilla and Volcano thusly: "The country was weird and Dantesque in the extreme. Leagues upon leagues of wild volcanic debris and weary stretches of arid alkali plains monopolized the landscape."[8]

About 5 miles south of Volcano was the tiny village of Skarda, which had a post office from 1922–1942. Skarda was named after a local rancher. "Homesteaders arriving after WWI made up most of the population, but dry years and hard times eventually led to Skarda's abandonment."[9] The D&RG had no facilities at Skarda, but the mail was picked up by the train and transported south to Tres Piedras.

Farther south, the Chili Line came to the desolate sounding place of No Agua, which is Spanish for "no water." The landscape may indeed have been desolate, but there was water here. A 50-foot-deep well provided water which was pumped up by a 51-foot-high Haliday windmill. At No Agua, the structures were on the east side of the tracks and consisted of the windmill, water tank, section house, bunk house, and coal bin. *The WPA Guide To 1930s New Mexico* also stated that No Agua had "a store, small church, and two or three houses. No Agua Mountain … looms against the horizon."[10] No Agua was 27.6 miles south of Antonito and the siding there could accommodate twenty-two cars.

The first town of any size approached by the Chili Line after Antonito was Tres Piedras, 34.7 miles to the south. Tres Piedras is Spanish for "three rocks," and that refers to the three large granite outcroppings on the west edge of town. Tres Piedras

K-28 no. 473 is headed south past San Antonio Mountain with a gondola car, RPO car, and a passenger car. This short consist was typical of the later years on the Chili Line. (*Friends, Dorman Collection, RD085-020*)

This contemporary view duplicates the historic view with the Chili Line grade and San Antonio Mountain. (*Photograph by the author*)

K-28 no. 471 is shown at Palmilla with snow on the plow and locomotive on this snowy day of March 1941. (*Friends, Dorman Collection, RD086-039*)

K-28 no. 471 is seen near Palmilla on a distinctly warmer day than the previous photo. Behind the locomotive are a flat car, Conoco oil car, boxcar, RPO car, and passenger coach. (*Friends, Dorman Collection, RDS071-226*)

Antonito to Tres Piedras

This is the volcanic landscape between Palmilla and Volcano, with volcanic rocks in the middle, and extinct volcano Ute Mountain in the distance before the snow-capped Sangre de Cristo Mountains. (*Photograph by the author*)

K-28 no. 475, heading south, pulls into No Agua on July 23, 1940, with a mixed train. (*Friends, Dorman Collection, by Gerald Best, RDS077-088*)

No Agua as seen from the rear of the passenger coach as viewed by photographer Robert W. Richardson on July 2, 1941. The No Agua Peaks are to the north, and just beyond the buildings at the foot of the peaks is where the perlite mine is located today. (*Friends, Dorman Collection, RD086-074*)

A mixed Chili Line train heading south at No Agua. (*Friends, Dorman Collection, RD086-054*)

K-28 no. 475 heading south toward Tres Piedras. The landscape between No Agua and Tres Piedras is fairly flat. (*Friends, Dorman Collection, RDS077-029*)

was settled in 1879 with the prospect of the railroad on the horizon. While the view east of Tres Piedras is flat, desolate desert scrubland (except for the majestic Sangre de Cristo Mountains looming some 30 miles across the plains), the landscape west consists of rolling mountains covered with timber. Cattle were raised and timber was harvested in these mountains.

The D&RG built two narrow gauge short-line railroads branching westward off the Chili Line, each named "Stewart Junction," one just north of town and one just south. They served the lumber operations to the west. Stewart Junction No. 1 was 4 miles south of Tres Piedras and operated from 1888 to 1890. Stewart Junction No. 2 was 3 miles north of Tres Piedras and operated from 1890 to 1892. In Tres Piedras, the water tank was two-tenths of a mile north of the depot, and the water came from a well with a Haliday windmill. The depot was a four-room frame structure. There was also a coal shed and platform. These structures were on the east side of the tracks. The siding could accommodate twenty-eight cars.

A trackside view looking east at the Tres Piedras depot, *c.* 1900. (*Friends, Dorman Collection, RD086-068*)

C-16 no. 205 with a freight train at Tres Piedras taken by R. E. Marsh, July 1911. (*Friends, Dorman Collection, RD086-059*)

A seven-car mixed Chili Line train heading north at the Tres Piedras depot. (*Friends, Dorman Collection, RD086-080*)

The south end of the Tres Piedras depot as viewed from the passenger coach. The conductor's "bay window" is visible at the side of the coach. Since Chili Line trains seldom (if ever) had a caboose from which the conductor could view the train ahead of him, Chili Line conductors used this bay window to view the train. (*Friends, Dorman Collection, RD086-083*)

Tracking the Chili Line Railroad to Santa Fe

Above: A snowy scene just south of the Tres Piedras depot on March 3, 1941. The train is halted, waiting for the rotary snowplow to arrive from Antonito after clearing the line, so this train could proceed on its northbound journey. (*Friends, Dorman Collection, RDS077-080*)

Left: Trackside view of the Tres Piedras water tank, taken by Donald Rogers on May 24, 1939. (*Friends, Dorman Collection, RD086-063*)

The Tres Piedras water tank is falling into disrepair in this contemporary view. Notice San Antonio Mountain to the north. (*Photograph by the author*)

TRACKING THE CHILI LINE TODAY

US Highway 285 essentially follows the Chili Line grade from Antonito south to Tres Piedras. It is 6 miles from Antonito to the New Mexico border. Just south of the border, it is fairly easy to spot the grade off to the left (east). The grade is generally marked by abundant sagebrush plants growing in a straight line or curving where the railroad grade curved. Some 5 miles south of the border would have been the location of Palmilla. At mile marker 401, a left turn takes the traveler on a road out to the Taos Plateau, which is marked with a Rio Grande del Norte National Monument sign and map. Just past the sign, the dirt road crosses the grade and it can easily be seen to the north and south. Continuing back south on US285, at about 7.5 miles south of the border, notice a right turn (west) to Forest Road no. 118. Immediately east at this point, the Chili Line grade is quite noticeable and there is a cut through the grade where there was a small trestle or culvert pipe. In the distance to the east is Ute Mountain, another extinct volcano.

Between mile markers 398 and 397, US285 crosses the Chili Line grade. The grade is quite noticeable on the west side of the highway. The summit to the west is where Volcano siding would have been. At the intersection with Forest Road no. 87, turn right (west) and drive about one mile on this good gravel road. Look to your right (north) where there is a very noticeable curving line of sagebrush. This is the Chili Line grade. If you wish to hike it, park at the edge of the road and head north. The grade is elevated, and in about three-quarters of a mile, remains of railroad ties can be seen. A

Tracking the Chili Line Railroad to Santa Fe

The V-shape in the center foreground is where there was probably a small trestle or culvert pipe connecting the two sections of grade on the left and right. Volcanic rock is scattered across the plateau and the Sangre de Cristo Mountains rise to the east beyond Ute Mountain. (*Photograph by the author*)

hike of about 2 miles on the grade takes you back to US285, where your ride could pick you up if so desired.

Traveling south on US285 once more, you will come to the Red Hill mine (lava products) on the right (west) and the No Agua Perlite Mine on the left (east). The Chili Line No Agua section house and bunk house would have been just south of the perlite mine entrance.

South of No Agua, the Chili Line grade is quite a bit east of the highway, and only glimpses of it can occasionally be seen between here and Tres Piedras. As you enter Tres Piedras, look on the left (east) side down in a valley and you will see the remains of the water tank.

The Chili Line grade is clearly marked by the elevated section of sagebrush at the center which then curves off to the right at the base of San Antonio Mountain. (*Photograph by the author*)

Perlite mining is occurring at the base of the No Agua peaks. The Chili Line structures at No Agua would have been just south of here. (*Photograph by the author*)

The Chili Line Depot restaurant greets travelers on US285 as they enter Tres Piedras from the north. The building is an old dance hall and saloon, but it was not the depot, which has long since vanished. (*Photograph by the author*)

COLORFUL CHARACTERS OF THE CHILI LINE: ALDO LEOPOLD

Carson National Forest has always played an important role in the lives of citizens of Tres Piedras who relied on grazing cattle and sheep in the forest and cutting timber for buildings. It was created in 1908 by combining the Taos Forest Reserve with the Jemez Forest Reserve. It stretches from north of Española all the way to the Colorado border. Taos artist Bert Geer Phillips worked as a forest ranger in the Taos Forest Reserve in 1907, when his eyesight was failing due to the strain of painting. A year in the forest helped clear up his vision, and when the combined forest was created in 1908, he proposed the name of Kit Carson National Forest in honor of his boyhood hero. This suggestion was enthusiastically approved by the Forest Service, and Carson National Forest was born.

At first, the headquarters for the Carson National Forest was in Antonito, but in 1912, it was moved to Tres Piedras through the efforts of a young forester named Aldo Leopold. Aldo Leopold is known today as a founder of the conservation movement and the wilderness system in America.

Leopold was born in 1887 in Burlington, Iowa. He attended the Yale Forestry School and graduated in 1909. One of his first assignments was working in the Apache National Forest in Arizona. His good work was soon noticed by his supervisor who called him up to Albuquerque to discuss promotion possibilities. Upon entering a drugstore there, they both noticed a pair of very attractive Hispanic sisters who were visiting from Santa Fe. It was virtually love at first sight for Aldo as he chatted with Estella Luna Otero Bergere. His supervisor noted the budding romance and transferred Leopold to Carson National Forest so that Aldo could be closer to Estella. Aldo was stationed at forest headquarters in Antonito in May 1911, but when he was promoted to superintendent of Carson National Forest in 1912, he moved the headquarters to Tres Piedras to be even closer to Estella. Here is where the Chili Line comes in. Aldo often rode the train to Santa Fe to visit Estella. He described the Chili Line as "slower'n a burro and just as sorry."[11] Other old-timers added that D&RG stood for "Dangerous and Rough Going."[12] At any rate, it was the Chili Line that took Aldo to Santa Fe to propose to Estella. They were later married in October 1912.

With wedding plans on the horizon, Leopold designed and built the cabin in Tres Piedras in which he and his bride-to-be would live, and which would serve as the headquarters building for Carson National Forest.

> [The Forest Service] approved $650 so Leopold could build new supervisor's quarters in Tres Piedras. He designed the cabin himself—a 1 ½ story bungalow set back against one of the village's massive granite outcrops and facing east toward the Sangre de Cristo Mountains.[13]

The cabin had a kitchen, bathroom, bedroom, library, and dining room on the first floor.

Aldo Leopold was faced with a big task as supervisor of Carson National Forest. "By 1900, the Upper Rio Grande may have been the most heavily grazed watershed in the country with 220,000 cattle and 1.7 million sheep."[14] Aldo's job was to bring rules and regulations to these abused and overgrazed forest lands. He rode a horse and packed a rifle as he met with hostile ranchers throughout his domain. Thus, it is a bit ironic that

This photo taken in 1946 by R. King shows the Tres Piedras ranger station established by Aldo Leopold. Aldo's home is in the center. (*USDA Forest Service*)

The scene in 2019 looks the same as in 1946 thanks to the restoration efforts of volunteers and staff of the Forest Service. Aldo Leopold's home in the center now serves as a summer retreat for writers. (*Photograph by the author*)

Aldo and Estella Leopold walk along the Chili Line tracks in Santa Fe, c. 1912. (*Aldo Leopold Foundation & University of Wisconsin Library, leo0078*)

his bride Estella Luna Otero Bergere was "heiress to one of the great sheep empires in the West," founded by her grandfathers, Don Jose Luna and Don Jose Otero.[15]

In April 1913, Estella was pregnant and took the Chili Line to Santa Fe to be with her family, while Aldo was riding the range in the far northwest part of the forest. Aldo fell desperately ill and barely made it back to Tres Piedras. Encouraged by a friend to see a doctor in Santa Fe, he also hopped aboard the Chili Line and headed south. In Santa Fe, it was discovered that he had Bright's disease, a serious kidney ailment that would soon have taken his life had it not been discovered and treated. He spent the next eighteen months resting and recovering. His and Estella's time at Tres Piedras was over.

Volunteers and the Forest Service restored the Leopold cabin at Tres Piedras in 2006. Since 2012, it has served as the home of the Aldo and Estella Leopold Writing Residency Program, which uses the cabin as a retreat for writers and artists to reflect on and write about the relevance of Aldo Leopold's ideas to twenty-first century cultural and environmental issues. Residents can stay in the cabin for up to one month during the months of May to October. The program is co-sponsored by the Forest Service and the Aldo Leopold Foundation, which was founded in 1982 "with a mission to foster the land ethic through the legacy of Aldo Leopold, awakening an ecological conscience in people throughout the world."[16]

Writers in the Residency Program today might well be inspired by Aldo's words in an autumn letter he wrote to Estella from Tres Piedras on September 25, 1911, reflecting his great love for this area:

> San Antone Mountain was a great glory of bronze and gold- and the Taos Mountains—60 continuous miles of main range under the eastern sky—ablaze with great masses of orange and crimson. I can hardly tell you what a blessed peace I find in my Sundays in the hills—I wouldn't be able to get along without them—something would break, I know.[17]

3

TRES PIEDRAS TO TAOS JUNCTION

With track laying completed to Tres Piedras on July 18, 1880, the Chili Line proceeded south to its next stop at a water tank 7 miles away. This was known as Servilleta Tank, with the little settlement of Servilleta another 2.5 miles south from there. Servilleta means "napkin" in Spanish, the name implying the flat plain upon which Servilleta was located. Servilleta had a post office from 1913 to 1949. The rail siding at Servilleta was 836 feet long and could accommodate twenty cars. Besides a bunk house and section house, Servilleta had two 9 × 35-foot freight car bodies, one of which served as a depot and the other as storage space.

Servilleta served as the original stagecoach stop for the town of Taos, nearly 25 miles to the east. In 1899, Oscar Berninghaus, a young artist from St. Louis, was hired by the D&RG "to sketch and produce watercolors of the mountain scenery, mining camps, people and villages," for the railroad's advertising brochures.[1] Berninghaus rode the standard gauge D&RG from Denver to Antonito, where he then embarked upon the Chili Line. Berninghaus noted that the ride was slow, with frequent stops, which gave him ample opportunity to make sketches along the way. He then had what every rail fan dreams of—a locomotive cab ride. Only it was not a cab ride—it was a ride on top of one of the freight cars. Here is his description:

> The train crew, taking an interest in me and what I was doing, suggested I might ride the top of the freight car that I might better see the country as we rolled along, but not before I was securely strapped to the brakeman's iron guard rail that ran atop each car … As we stopped (at) Servilleta … the brakeman pointed out a certain mountain lying toward the east; this he called Taos Mountain.[2]

Intrigued by this mountain, Berninghaus departed the train at Servilleta so that he could visit Taos. He hopped aboard John Dunn's stagecoach and vividly described the ride:

> I started on a twenty-five mile wagon trek over what was comparatively a goat trail. The trip took 10 hours, and the wild expanse of mountain and desert, the curious coyotes and pronged-horned antelopes that trotted along behind the coach or stood

Two freight car bodies on the left served as depot and storage at Servilleta, in this wintry view taken by John W. Maxwell in March 1941. (*Friends, Dorman Collection, RDS077-086*)

This summer view at Servilleta shows the same two freight car bodies on the left, and the same tree configuration on the right. In the middle, K-28 no. 473 is stopped in this photograph taken by Robert W. Richardson on July 2, 1941. (*Friends, Dorman Collection, RD086-051*)

To the right of the boxcar "depot" at Servilleta, K-28 no. 475 is switching cars on the siding, while the main Chili Line consist awaits on the left, *c.* 1930s. (*Friends, Dorman Collection, RD086-079*)

Arthur Rothstein took this photo of what Oscar Berninghaus referred to as the "massive piles of the pueblo," at Taos Pueblo in April 1936. The view was probably not much different from that witnessed by Berninghaus in 1899. (*Library of Congress, LC-USF34-002937-D*)

close-by while the conveyance passed, delighted me as did the little adobe town (Taos) and the massive piles of the pueblo.[3]

Berninghaus was so enchanted by Taos and the pueblo that he returned to visit there every summer between 1900 and 1925. He was one of the six founders of the Taos Society of Artists in 1915. He finally settled for good in Taos in 1925, where he died in 1952. That first ride on top of the freight car on the Chili Line was still firmly etched in his memory in 1950 when he wrote about that trip.

The next stop south on the Chili Line is Taos Junction, 11.84 miles from Servilleta. While the landscape here is still relatively flat, the Sangre de Cristo Mountains rise majestically to the east. *The WPA Guide To 1930s New Mexico* described the slow pace of the Chili Line:

> [It was] the equivalent of a walking tour by train ... The trip ... is leisurely enough to give the illusion of walking through fields of grasses and wild flowers, and this is heightened by growth right up to the tracks. There are panoramas not seen from any highway in the State.[4]

Until 1914, Taos Junction was known as "Caliente," named for the Ojo Caliente hot springs about 12 miles to the south. The D&RG built a 16.7-mile narrow-gauge branch line from Caliente to Ojo Caliente and La Madera in 1914 to serve the Hallack and Howard lumber operation at La Madera. At that time, the name Caliente was changed to Taos Junction, signifying this junction to the branch line. It was a bit confusing because it was not a railroad junction to the town of Taos. Prior to the creation of the La Madera branch, travelers could reach Ojo Caliente Hot Springs by departing the Chili Line at Barranca (8.6 miles south of Taos Junction) and traveling by stagecoach from there to the hot springs. The December 3, 1900 *Santa Fe New Mexican* newspaper noted:

> The temperature of these springs is from 90 to 123 degrees. The gases are carbonic. Altitude 6,000 feet. Climate very dry and delightful the year around. There is now a commodious hotel for the convenience of invalids and tourists.[5]

The La Madera branch from Taos Junction owes its origin to a 1913 U.S. Forest Service auction, which advertised 117 million board feet of timber for sale in Carson National Forest north and northwest of La Madera, a small village about 15 miles west of the Chili Line. The Hallack and Howard Lumber Company won the auction with a bid of $2.50 per thousand board feet, and they planned to begin logging operations by July 1, 1915.[6] Hallack and Howard realized that they needed a railroad to get the harvested timber out to the Chili Line, so they negotiated a contract with the D&RG to build the line. Construction was completed by November 15, 1914, and the first timber was shipped out within a few days, well ahead of the 1915 planned date. With the extra trains from the lumber operation, as well as the daily runs of the Chili Line trains, the D&RG recognized the need for a larger depot at Taos Junction. "In the fall of 1915 ... a new depot was constructed to replace the former box car used for station purposes."[7] The new depot was a two-story structure, with living quarters for the station master provided in the upper story. The depot measured 20 × 32 feet with an additional wing of 20 × 64 feet. There was a bunk house in a freight car body which was 9 × 32 feet. The siding could accommodate twenty-eight cars.

Tres Piedras to Taos Junction

The two-story depot at Taos Junction was completed in 1915 after the branch line to La Madera was constructed. This was the only two-story depot on the Chili Line. This view was captured by John W. Maxwell on August 30, 1941. (*Friends, Dorman Collection, RD086-005*)

John W. Maxwell also took this photo at Taos Junction on August 30, 1941, with K-28 no. 473 heading up a mixed consist with a passenger coach on the rear. The gabled roof and the chimney of the depot can be seen just above the freight cars. (*Friends, Dorman Collection, RD086-002*)

E. J. Foley, Jr., shot this photo of Chili Line passenger coaches 320 and 306 with RPO no. 63, with snow piled in the background at Taos Junction, March 12, 1941. (*Friends, Dorman Collection, RDS077-102*)

Another Foley photograph from March 12, 1941, shows a taxi cab waiting for passengers at the Taos Junction depot. The lettering on the side of the taxi announced the route of the taxi traveling from Abiquiu to El Rito to Ojo Caliente to Taos Junction to Taos. (*Friends, Dorman Collection, RDS071-223*)

Tres Piedras to Taos Junction

K-28 no. 471 is headed southbound with a mixed consist in the 1930s at Taos Junction. Water car no. 0459 is parked next to the depot on a siding. The D&RG provided this water car for farmers and ranchers in this dry area. (*Friends, Dorman Collection, RD086-004*)

Two ranchers are preparing to draw water from water car no. 0459 at Taos Junction. Their water barrel is seen on the cart behind the horse. Ranchers could purchase this water from the D&RG. (*Friends, Dorman Collection, RD086-006*)

There was actually very little settlement at Taos Junction until 1921 when nearby lands were opened for settlement by the federal government to service men from World War One. The veterans then often sold their land priority to other settlers. One such settler was Gordon Watts, who noted that a few years later there were "two stores, a hotel, a recreation hall, gas tanks and a school," at Taos Junction.[8] This is a very dry area, however, and "To ease water shortages, the D & RGW hauled water in tank cars from San Antonio Creek south of Antonito. The settlers were able to purchase drinking water at 25 cents for a 50 gallon barrel."[9] Photographs of Taos Junction often show a water tank car parked at the depot. The Taos Junction settlement was largely abandoned by the end of 1935 when the federal government bought back nearly all the land as it was "submarginal" due to the lack of water. With the abandonment of the Chili Line in 1941, there was simply no more reason for a settlement to exist at Taos Junction.

The timber mill at La Madera "had a significant impact on the entire area. Employment was provided for about 300 men for 11 years, and about $1,000,000 in freight revenue (helped) sustain Chili Line operations."[10] In addition to the D&RG line to La Madera, "about 1922 another line was built by Hallack and Howard, this one to Vallecitos."[11] Vallecitos was a small village nine miles north of La Madera, in another heavily forested area. The villages of La Madera and Vallecitos still exist today. In addition to the sawmill at La Madera was the following:

> Hallack and Howard also operated a planing mill and box factory ... Boxes were shipped 'knocked down' to the San Luis Valley (in Colorado) for vegetables, and to Lamar, also in Colorado, for packing Rocky Ford canteloupe.[12]

By the summer of 1926, most of the easily harvested timber in the La Madera and Vallecitos areas had been cut, and logging operations had wound down. In La Madera, the sawmill was sold and dismantled in April 1927. The D&RG line to La Madera held on a bit longer though, before it was abandoned and dismantled in June, 1931.[13] The entire Chili Line operation suffered with this loss of logging revenue, and it helped bring about eventual abandonment of the line in 1941.

This view from 1920 shows the extent of the La Madera sawmill and lumber yard. The timber cutting operation and sawmill employed some 300 men at peak operation for the Hallack and Howard Lumber Company. (*U.S.D.A. Forest Service*)

The sawmill is long gone at La Madera, but evidence of better times in the past is shown in this contemporary view of a building with a sign marking the "La Madera Mercantile." Note the steep mountain slopes in the background. (*Photograph by the author*)

Vallecitos still looks like a prosperous village in this 1939 view by W. H. Shaffer of the Forest Service. Hallack and Howard built a railroad spur to Vallecitos in 1922, and there was a sawmill at Vallecitos. By 1932, however, all operations had ceased. (*U.S.D.A. Forest Service*)

There are still quite a number of buildings in Vallecitos today, but many are falling down. The little village appears like a ghost town, with only a few residents. (*Photograph by the author*)

A log-loader near La Madera hoists logs onto flat cars which will then be pulled out by the railroad, c. 1923. (*U.S.D.A. Forest Service*)

A loaded log train is transported out of the forest by a steam locomotive on the La Madera branch headed for Taos Junction in this 1923 photograph by W. J. Perry. (*U.S.D.A. Forest Service*)

TRACKING THE CHILI LINE TODAY

From Tres Piedras to Taos Junction, US Highway 285 essentially parallels the Chili Line grade which was just to the east of the highway. This grade is largely invisible today though, due to the dense piñon and juniper forest growth. Taos Junction is marked by the remains of the old Taos Junction Cafe on the west side of the highway, about one-eighth of a mile before the intersection with New Mexico Highway 567. It was at this intersection that the Taos Junction depot stood, and the Chili Line grade can be traced south from here along Forest Road 557.

From Taos Junction, continue south on US285 to visit the villages of La Madera and Vallecitos. In approximately 8 miles, turn right on New Mexico Highway 111. Follow this highway for 5 miles to La Madera. Then continue on for 9 more miles to Vallecitos, turning left on NM576 into the village of Vallecitos. With its collapsing buildings, Vallecitos has the appearance of a ghost town, but there are still several residents living here, so please respect private property. There are no visible railroad grades on the drive to La Madera and Vallecitos, but the drive goes through beautiful mountain scenery, heavily forested with juniper and piñon trees, and ponderosa pines at higher elevations toward Vallecitos. The forest appears to have recovered nicely from the logging activities of the 1920s when clear-cutting was prohibited in this area.

The building which housed Maria's Taos Junction Cafe Bar is still standing, but the business is long gone. The building is about 1/8 of a mile north of the junction of US285 and NM567. (*Photograph by the author*)

COLORFUL CHARACTERS OF THE CHILI LINE: "LONG" JOHN DUNN

"Long" John Dunn arrived in Taos in 1889 from Texas. He earned the nickname "Long" because of his height (6 feet 4 inches) and slim build. Noting that Taos had no link to the Chili Line railroad some 25 miles to the west, Long John decided that a perfect business opportunity for him would be to operate a stage line out to Servilleta. To accomplish this, Dunn purchased a bridge across the Rio Grande from a Mr. Meyers for $2,200.[14] The bridge linked Servilleta on the west with Arroyo Hondo on the east. From Arroyo Hondo, the stage line proceeded south to Taos. With the purchase of the bridge, Dunn was then in complete control of the lucrative business of hauling passengers from the train depot at Servilleta to Taos. He also charged tolls to anyone else crossing the bridge, and a toll for each cow and sheep that crossed. He then built a hotel where the Rio Hondo emptied into the Rio Grande. Viewing the completed hotel, Long John "declared proudly:"

> "It looked like a castle setting up there against those rocky bluffs." He always delayed his last stagecoach so the passengers would have to stay all night in this fine road-ranch. "It was really too much of a haul to make the trip in one day, anyway," John said. They were given clean rooms, and John fed them well.[15]

Long John developed quite a reputation for himself in Taos. He established hotels, saloons, and gambling joints. His biographer, Max Evans, noted that John "was top-notch in everything he did. John was one of the best gunfighters, gamblers, bronc riders, ropers, stagecoach drivers, trail-herd drivers, saloonkeepers, outlaws and, ironically, hardheaded businessmen."[16] With all of his hard-earned cash, Long John built a ten-room adobe house on Bent Street in Taos, which still stands today with several retail businesses operating therein. The bridge west of Arroyo Hondo is still known as the John Dunn Bridge, although the original log bridge has long since been replaced by the present steel truss bridge.

When the La Madera branch line was built off the Chili Line in 1914, "the D & RG gave materials for a new bridge across the Rio Grande to be constructed at a point above Pilar. From then all passengers detraining for Taos used Taos Junction."[16] Keeping up with the times, John Dunn then switched his stage line operation from Servilleta to Taos Junction. Travelers today can still cross this bridge by driving east from Taos Junction on New Mexico Highway 567 to Carson, and then turning south at the intersection with NM570 to Pilar. The descent down the gravel road to the Rio Grande is quite steep and narrow, and four-wheel drive is highly recommended. Pavement resumes after the road crosses the bridge. From Pilar, travelers can then take NM68 north to Taos.

This c. 1915 view shows the Howe truss bridge over the Rio Grande north of Pilar. It is about 10 miles east of Taos Junction, and is commonly called the Taos Junction Bridge, though the Chili Line never descended to this point of the Rio Grande. This was a bridge for wagons, stagecoaches, and later, for automobiles. (*Friends, Dorman Collection, RD087-094*)

This is a contemporary view of the Taos Junction Bridge. It can be reached by taking NM567 east from Taos Junction, and then descending into the Rio Grande canyon on NM570. This is now part of the Rio Grande del Norte National Monument, and there are campgrounds along the river. (*Photograph by the author*)

4

TAOS JUNCTION TO EMBUDO

Caliente (Taos Junction) presented a dilemma for the tracklayers of the Chili Line. The question was how to get off the plateau and into the Rio Grande Gorge some 1,500 feet below. One route would have led down to Pilar (reached today by NM567 and NM570). The other route would descend precipitously from Barranca into Comanche Canyon and then on to Embudo. The Comanche Canyon route involved a rugged 4-percent grade, but it would cost $200,000 to $300,000 less, so the D&RG chose Comanche Canyon. As mentioned before, this would necessitate double-heading locomotives for long freight trains up the grade from Embudo in the early days before more powerful locomotives came along. At the top of Comanche Canyon, a wye was necessary at Barranca, so that the extra locomotive could turn and head back to Embudo. At Embudo, there was a turntable so that locomotives could be turned around for the next uphill journey.

From Taos Junction, the elevation drops 375 feet on a gentle 8.5-mile grade to Barranca. "Barranca" is Spanish for ravine or gorge, and that is exactly what the builders faced as they dropped off the cliff heading south. The railhead reached the foot of Comanche Canyon on November 19, 1880, and from there, it was an easier journey along the Rio Grande to Embudo. Barranca had a 1,328-foot-long siding, which could accommodate thirty-four cars. It had a section house, a bunk house, and a tool house, as well as the wye for turning the locos. Barranca had stagecoach service to Ojo Caliente Hot Springs, as well as to Glen Woody directly below on the Rio Grande with connections from there to Taos by stage. Glen Woody was named for its founder who had discovered some quartzite deposits on the east side of the Rio Grande. He built a cable car to bring his ore deposits down from the mine and across the river to the west side where his mill was located. The processed ore was then hauled by mule up to the top of the mesa to Barranca, where it was shipped out by the Chili Line.

Extraction costs were excessive and Glen Woody turned to a stage coach line to make his fortune … In 1910 a traveler attempting to reach Taos would leave Santa Fe on the Chili Line at 10:15 a.m., have lunch at Embudo, and arrive at Barranca at 2:25 p.m. Leaving Barranca behind four horses would, with luck, place the weary traveler

A Chili Line train heads up the steep Comanche Canyon grade with a consist of a gondola, RPO, and passenger coach, c. 1940. (*Friends, Dorman Collection, RDS077-139*)

K-28 no. 475 is headed down the Comanche Canyon grade with a string of freight cars, c. 1940. Before descending from Barranca, it was essential to set the double retainer brakes. (*Friends, Dorman Collection, RDS071-016*)

in Taos at 6:30 p.m. Woody's line folded in the 'teens with the advent of Taos Junction and the automobile.[1]

The approximate location of the Glen Woody operation was where the hanging bridge can be seen across the Rio Grande along NM68 today, about 3 miles north of Rinconada.

Comanche Canyon was undoubtedly named after the Comanche Indians, who raided settlements along the Rio Grande in the mid-1700s. The Ranchos de Taos area, just south of Taos, was settled by Spanish ranchers in the 1730s, and by 1760, a village had arisen. It "immediately came under severe attacks by Comanche warriors, and residents fled for their lives to Taos Pueblo, where they stayed until 1779."[2] By the time the Chili Line was built in 1880, the Comanche were long gone, but the name remained. The chief obstacle for the Chili Line builders was not Indians, but the 4-percent grade of Comanche Canyon. "Here on the steep grades in the Canon, 30 pound steel rails were to be laid instead of iron, the better to survive hard usage on the curves and the grade."[3] Still, the steep grade presented a challenge to locomotive engineers and wrecks did occur. One notorious wreck on July 17, 1929, killed the engineer and the fireman.

"*Embudo*" is Spanish for "funnel," and here, the Rio Grande does funnel its way from the narrow northern gorge to the broader valley south of Embudo at Velarde. The original village of Embudo was located where the Rio Embudo empties into the Rio Grande. This was about 2 miles north of present-day Embudo, and when the D&RG named its station Embudo, it "caused great confusion and led to the renaming of the old Embudo settlement to Dixon, in honor of the area's first school teacher Collin Dixon."[4]

Embudo was one of the largest operations on the Chili Line. In addition to its 50-foot turntable, it had a siding of 1,311 feet, which would accommodate thirty-three cars. The depot was 27 × 33 feet. There was a water tank, freight house, sand house, bunk house, coal house and coal chute. The "eating house" was 16 × 43 feet with a 16 × 28-foot wing. Since trains took on both water and coal at Embudo, passengers had a chance to detrain and head to the restaurant, which operated from 1900–1930.

Henry Wallace became station master at Embudo in 1912. Wallace had tuberculosis and was given only a few years to live. However, the dry, sunny New Mexico climate enabled him to survive for twenty-five more years. With only two trains per day passing through Embudo (one northbound from Santa Fe; one southbound from Antonito), Wallace had a lot of time on his hands. He filled the empty hours by cementing cobblestones to the outer walls of the station. When the D&RG management observed what he was doing, they decided to help him out. "Carloads of cement and rock were sent to him to continue the work."[5] So, stone by stone, Wallace covered the walls of the station. "When the line was being abandoned in 1941, the railroad discovered his stonework to be so substantial that wrecking costs would have far exceeded the salvage."[6] Thus the station still stands today and is a private residence.

Due to increasing financial losses, the D&RG closed the Embudo station in 1934. A watchman and the section crew continued to protect the property. Henry Wallace was transferred to Santa Fe where he became the station agent, and he served there until his death on March 1, 1937.

K-28 no. 473 glides along the Rio Grande to Embudo after surviving the descent from Barranca, c. 1940. (*Friends, Dorman Collection, RDS077-060*)

A relief train is on site at Barranca Hill as it offers assistance for the horrific wreck which occurred on July 17, 1929. The engineer and the fireman were killed in the wreck. (*Friends, Dorman Collection, RD086-013*)

This was the Embudo station in about 1910 as passengers wait to depart on the Chili Line. This was before station agent Henry Wallace arrived in 1912 and began adding the cobblestone veneer to the exterior walls. (*Friends, Dorman Collection, RD087-010*)

In this *c.* 1930s view of the Embudo station, the cobblestone veneer has been completed by Henry Wallace. Some well-dressed men pass by the station on an inspection tour atop a "speeder" (a gasoline-powered flat car). (*Friends, Dorman Collection, RDS073-052*)

Southbound K-28 no. 473 is taking on water at the Embudo tank on July 2, 1941, as viewed by passenger Robert W. Richardson, founder of the Colorado Railroad Museum. (*Friends, Dorman Collection, RD087-025*)

K-28 no. 473 is southbound at Embudo station on July 2, 1941, in this photograph taken by Robert W. Richardson. (*Friends, Dorman Collection, RD087-012*)

K-28 no. 471 is hauling freight toward Embudo along the banks of the Rio Grande, *c.* 1940. (*Friends, Dorman Collection, RDS071-047*)

K-28 no. 471 has reached the Embudo water tank along with its freight consist, *c.* 1940. (*Friends, Dorman Collection, RDS071-210*)

K-28 no. 471 is pulling out of Embudo heading north along the Rio Grande. A portion of the water tank can be seen at the top right, and the bridge across the river to the highway can be seen near the top left. (*Friends, Dorman Collection, RDS077-049*)

TRACKING THE CHILI LINE TODAY

At the intersection of US285 and NM567, turn east and then immediately turn right to Forest Road 557. This road essentially follows the grade of the Chili Line to Barranca. The road is dirt and is deeply rutted from rain and snow. Four-wheel drive is necessary to navigate this road. As you drive along the road, there are obvious sections of railroad grade.

After approximately 8.5 miles, you will see where the road surface turns from brown to gray. This large gray surface area is where Barranca was located. On the east side of the road, a walk through the area reveals the foundations of a couple of buildings, as well as a cistern. If you use your imagination, you can envision the area where the wye was located, just to the northeast of the building foundations. Road 557 continues to the west, though it is much less well-defined. A left turn on Road 557F will take you a half-mile to the top of the mesa, where there is an incredible view of the Rio Grande valley below with the little village of Rinconada hugging the banks of the river. It's hard to imagine the Chili Line descending into this valley, but that is just what it did, as it began its descent just a bit to the west of this point into Comanche Canyon.

Since the road does not descend into Comanche Canyon, to continue tracking the Chili Line, you must return to Taos Junction and take NM567 to its junction point with NM 570. Turn right on to NM570 and the road immediately begins its steep,

Heading south on Forest Road 557, the embankments of the Chili Line railroad grade are readily visible in several areas. (*Photograph by the author*)

At the top of Forest Road 557F, this magnificent view to the southeast rewards the traveler for enduring the 8.5-mile bumpy ride to this point. The Rio Grande is seen in the mid-distance with the village of Rinconada, a bit farther to the right on the east side of the river. (*Photograph by the author*)

switchbacking descent into the Rio Grande gorge. Passenger cars can make this journey, but it is very steep and narrow, and four-wheel drive is highly recommended. Due to the extremely tight switchbacks, do not pull a trailer down this road. It is quickly apparent why the D&RG chose Comanche Canyon for the Chili Line's descent, rather than this convoluted path. Breathe a sigh of relief when you reach the Taos Junction bridge over the Rio Grande, and then enjoy the smooth, flat drive along the river south to Pilar. At Pilar, the road joins NM68. Turn right and drive on south to Embudo. At Embudo, turn right across the bridge, and park in the large parking area that is available. You can walk to the station and the water tank, but all of these structures are on private property now, so please take photographs from a distance. Back on NM68, another half-mile south will bring you to a pull-out along the riverbank. Here, there is a sign for the stream gauging station, which can still be seen across the river.

The Embudo water tank still stands at the base of the rugged cliffs of Comanche Canyon. (*Photograph by the author*)

Henry Wallace's cobblestone covered Embudo station still stands today. It is now a private residence. (*Photograph by the author*)

COLORFUL CHARACTERS OF THE CHILI LINE: JOHN WESLEY POWELL AND THE HYDROGRAPHERS OF EMBUDO CAMP

As amazing as it may seem, the Chili Line was directly responsible for the establishment of over 8,000 stream-gauging stations in the United States, and the very first was at Embudo. It was the Chili Line that brought in the personnel for the stream gauging station at Embudo in December 1888. This unlikely story owes its origin to explorer John Wesley Powell, who had rafted down the Colorado River in the Grand Canyon in 1869. Powell was an anthropologist and geologist, and during his explorations of the West, he realized that this arid land could not support the traditional family farm structure found in the agricultural areas of the eastern United States. West of the 100th meridian, there simply was not enough precipitation to support thousands of small farms. In 1878, Powell recommended to Congress an Irrigation Survey of the West to determine precipitation and stream flow in order to suggest land use patterns in the arid lands. This was not approved, but in 1881, Congress did approve the establishment of the U.S. Geological Survey, and Powell became its director.

In October 1888, Congress finally did approve Powell's Irrigation Survey, and it was instituted as a branch of the U.S. Geological Survey. The purpose of the survey was "mapping drainage basins, measuring stream flow (hydrography), and assessing potential sites for irrigation canals and reservoirs."[7] Recognizing the need to act upon the Congressional funding as soon as possible, Powell selected Clarence E. Dutton, an Army captain and geologist, to oversee the Irrigation Survey. Embudo, New Mexico, was chosen as the site to begin the survey. Several factors led to Embudo's selection: firstly, it was west of the 100th meridian, as required by Congress; secondly, the Congressional appropriation bill was passed on October 2, and to get underway immediately, no northern streams could be chosen because they freeze over in the winter; thirdly, "Convenient railroad transportation directly to the site was necessary to expedite the project.," and fourthly, the Rio Grande at Embudo was just about the right size for the intended study—it was not too wide or deep, with a winter depth of only 6–12 inches.[8]

Dutton remained in Washington D.C. to oversee the Irrigation Survey, and he appointed Frederick H. Newell (a Massachusetts Institute of Technology graduate with a specialty in geology) to direct the Embudo project. Hydrography was a relatively new science, and while Dutton had access to plenty of topographers and geologists, there was an acute shortage of hydrographers. So Dutton decided to establish a hydrology school at Embudo to train hydrographers. He selected fourteen young men, graduates in engineering and geology, from East Coast Universities, and sent them west to Embudo. The first group to arrive included Frederick Newell and several others.

> And so it was on the 9th of December in 1888 a group of at least eight young engineers, most of whom had recently been graduated from eastern colleges, stepped off a coach of the narrow gauge Denver and Rio Grande Railway at Embudo, New Mexico.[9]

Imagine the sights that greeted these young men—rugged mountain slopes covered in cactus and piñon trees, adobe homes, and Spanish-speaking inhabitants. It must have been quite a culture shock for them.

> It seems very likely that the group of young engineers ... spent their first night or two in the tiny railroad station because there were very few other suitable buildings in the area. But on December 10, 1888 ... a supply of tents, purchased from the Army especially for their use, and a supply of folding cots arrived.[10]

It seems however, that John Wesley Powell had underestimated the severity of winters at Embudo's elevation of 5,800 feet, and the students were soon freezing in their tents and cots. Most of them abandoned the cots, dug trenches inside the tents, and wrapped up in blankets and slept in the trenches for more warmth. Other students slept in a cave on the mountain side, with a fire for warmth. It was a new experience for all of them.

The students spent the winter practicing with the various water measurement instruments and trying out new techniques.

> The work required of the men consisted in practicing stream gauging by various methods, measuring the rise and fall of the stream from day to day, measuring the daily evaporation, and making observations with meteorological instruments.[11]

For the students, it was very much a process of trial and error. For example, the first attempt at crossing the river was accomplished by building a raft from four barrels

Embudo tent camp for the training of hydrographers, December 1888. Some of the students arrived on the Chili Line the day before the tents did, and so spent the night in the train station. (*United States Geological Survey, h2op0165*)

and stringing a rope across the river. One can imagine that this resulted in a few dunks in the river. To improve the process, a rowboat was acquired, and a steel cable was stretched across the river. Then the students could tie the rowboat to the overhead cable and steady the boat in midstream in order to take measurements. Progress from that point on was rapid, and the work was completed and the camp abandoned in April 1889. The stream gauging station established at Embudo by the students has remained in operation to the present day, though no readings were taken from 1904–1912. With the appointment of Henry Wallace as D&RG stationmaster in 1912, readings were taken by him:

> Mr. Wallace, (as) has been previously mentioned … veneered the railroad station with cobblestones. Inasmuch as he has been reported as having applied such stone facings onto 'every object in the vicinity that would hold still,' he might very well have had a hand in applying the same kind of a facing onto the new recording gage shelter."[12]

The stream gauging shelter, covered in cobblestones, can still be seen along the west bank of the Rio Grande today.

This photograph shows the Embudo stream gauging station, c. 1899. (*United States Geological Survey, h20p0012*)

Opposite: The cobble covered stream gauging structure can still be seen today on the west bank of the Rio Grande just south of Embudo. It is still in operation. (*Photograph by the author*)

5

EMBUDO TO ESPAÑOLA

The Chili Line was completed to Española on December 31, 1880. It could proceed no farther south due to the infamous "Treaty of Boston," which declared that the D&RG could not expand southward for a period of ten years. "The line was declared by the railroad company completed and in operation as of December 31, probably for accounting reasons."[1] However, stagecoach connections to Santa Fe did not begin until January 4, 1881.

> In the months that followed, an unattractive town was laid out a short distance south of San Juan (pueblo) and named Española, and this became the railroad's actual terminus. It was a raw and lawless place indeed.[2]

Artist Birge Harrison did not find conditions in Española much improved on his Chili Line trip in 1885.

> [Leaving Embudo] the train wandered on for twenty miles or so through sprouting fields of Indian corn and green peppers, and drew up at length at the nondescript collection of canvas tents and board shanties on a flat beside the river. This was Española.[3]

At least, there was a frame depot when Harrison arrived. A tent had served as the D&RG's depot in Española for the first two years.

As the Chili Line headed south from Embudo, it hugged the west bank of the Rio Grande, coming to Velarde, some 3 miles away. Here, a bridge was built across the Rio Grande to carry a spur line into the village of Velarde, which was on the east bank of the river. The spur was necessary to capture the rich produce of the Velarde valley, which included cherries, plums, melons, peaches, apples, tomatoes, and chili peppers. A large warehouse in Velarde was used to store the produce which the local farmers brought in. Then the Chili Line steamed in to bring out the produce to Santa Fe and Denver. The siding at Velarde could accommodate twenty-five cars. Velarde had a depot consisting of a 9 × 35-foot freight car body, a 5 × 33-foot cinder platform, and a 12 × 16-foot wood platform. In 1875, Matias Velarde founded this farming community, which was in a broad valley or basin east of the Rio Grande. The community was originally named La Joya (Spanish for "basin"), but later took on the family name of Velarde, when David Velarde became postmaster in 1885.[4]

Some 5 miles south of Velarde, the next stop for the Chili Line was Alcalde. This stop was named after the village on the east side of the Rio Grande. "Alcalde" is Spanish for magistrate, or judge, which is appropriate since Alcalde was the county seat for Rio Arriba County from 1860–1880. Alcalde had a siding capacity of twenty cars. It also had a bunk house, section house, and a water tank, which was later abandoned.

Some 6 miles south of Alcalde is the village of Chamita, which means "little Chama" in Spanish. It was named for the Rio Chama, which flows beside the village. The village is located slightly northwest of the Rio Chama's confluence with the Rio Grande. The Chili Line crossed the Rio Chama just before the confluence, remaining on the west side of the Rio Grande to Española. Chamita is adjacent to the site of San Gabriel, which became the first "capital" of New Mexico. In 1598, Spanish conquistador Don Juan de Oñate brought a contingent of about 600 settlers, soldiers, and priests to what he named "San Juan Pueblo" on the east bank of the Rio Grande. Facing hostility from the natives at San Juan, Oñate moved his group across the Rio Grande to the west bank, where they inhabited an abandoned native pueblo.

> As headquarters of the Spanish colonizers, San Gabriel thus was the first Spanish capital of New Mexico, but the distinction was short-lived; about 1610 the Spaniards moved to what is now Santa Fe, and San Gabriel was abandoned.[5]

The Chili Line's facilities at Chamita consisted of a siding with a capacity of twenty-nine cars, a freight car body depot, a car body freight house, a cinder platform, and a warehouse.

Española is about 5 miles south of Chamita on the west bank of the Rio Grande. When the Chili Line reached the end of its allotted length in 1880, there was no town of Española. There was a collection of small Hispanic villages on the east side of the Rio Grande. The railroad set up a tent village on the west bank and named it "Española." "*Español*" means the Spanish language, so "*Española*" means Spaniard, or Spanish speaker.

> An 1882 business directory of New Mexico listed Española's population as 150 persons, and as late as the 1950s Española was officially only the small part of the present community that was on the west side of the Rio Grande, while Riverside and the other communities were separate settlements on the east side. Now all have come within the city limits of Española.[6]

When the frame depot was completed in Española in 1882, Samuel Stauffer McBride was appointed as the first station agent. He was a Union veteran of the Civil War, and he and his wife, Ella, had eight children. The depot did have living quarters for the station agent, but it must have been a tight fit for a family of ten. McBride served as station agent until his retirement in 1918 at the age of seventy-one. He must have had a good sense of humor in order to deal with the wildness of Española and the various characters who traveled through. One new lady resident was waiting for the train to Santa Fe after 1887:

This photo by Robert W. Richardson taken on July 2, 1941, looks east across the Rio Grande toward Velarde. The Chili Line bridge is already in a state of disrepair and the entire line will be abandoned on September 1, 1941. This spur into Velarde allowed the Chili Line to pick up the many fruits and vegetables produced by farmers in the Velarde valley. (*Friends, Dorman Collection, RD087-028*)

K-28 no. 473 is southbound at Alcalde with a mixed freight on July 3, 1941, in this photograph by Robert W. Richardson. (*Friends, Dorman Collection, RD087-029*)

K-28 no. 475 is southbound near Chamita with one freight car, R.P.O. car, and a passenger coach in this 1930s-era photograph taken by Gerald M. Best. (*Friends, Dorman Collection, RDS073-038*)

This scene at San Juan Pueblo (now known by its native name of Ohkay Owingeh) was captured by famed photographer Edward S. Curtis, *c.* 1904. This view cannot be duplicated today because a row of homes has been built across the street at approximately the spot where the native women are standing. (*Library of Congress, LC-USZ62-56048*)

A contemporary view of Ohkay Owingeh from a block north of Edward Curtis' photo. The San Juan Caballero Church with its square bell tower was built of brick in 1912, replacing the church seen in the Curtis photograph. (*Photograph by the author*)

[She] asked when the train was due, and was told that it would be there soon, the station agent adding, "You'll know it's coming when you see the engineer's dog running down the track ahead of it."[7]

There were no doctors in Española in the early years, so McBride provided healthcare to the residents with assistance of the Dr. Humphrey's Homeopathic Medicine Chest. There was no mention of how the patients fared after their treatment. McBride was also known for his friendliness in greeting newcomers and helping them to find lodging. In 1885, artist Birge Harrison and his wife arrived in Española and found that there was no hotel or boardinghouse. They had planned on staying in Española while Harrison painted the local scenery. Where, then, could they stay? Harrison noted:

> We at length found a man however, who consented to show us the only uninhabited domicile in town. His name was McBride and he was the station-master … McBride proved to be a jewel of a man. He not only let us have the house at a very moderate rent, but he supplied us with bedding, two chairs, some rugs, and a lamp.[8]

Due to the extreme lack of facilities in town, McBride allowed the depot to be used as a school and for Protestant church services, until 1890 when he and his son built an adobe Methodist Church. McBride served as station agent for thirty-six years until 1918, when he retired, and he died two years later at age seventy-three. Author John Gjevre paid tribute to McBride, noting that "Here was a railroad man in the noblest tradition, giving of himself freely to the Chili land people. They in turn loved him as one of their own."[9]

K-28 no. 473 is seen at the Española depot in this view by Robert W. Richardson on July 2, 1941. Mail loading is in progress on the R.P.O. car behind the tender. (*Friends, Dorman Collection, RD088-043*)

Heavy rains had weakened the roadbed, causing this derailment just north of the Española depot in 1921. Notice the large amount of standing water on each side of the track. (*Friends, Dorman Collection, RD088-017*)

Cleanup of the 1921 derailment is underway in Española. There are four cars on their sides with a load of sacks coming out of a boxcar roof. The standing water illustrates the great quantity of rain received. (*Friends, Dorman Collection, RD088-016*)

It was a quiet day at the Española depot on July 24, 1940, in this photograph taken by Gerald M. Best. Behind the depot at the building with the "DISPEN" sign, is where the first G.W. Bond & Bro. mercantile store was located in the 1880s. The street crossing the Chili Line tracks at the far end of the depot was Main Street, today called Paseo de Oñate. (*Friends, Dorman Collection, RDS071-204*)

A track gang poses with their section car in Española in 1916. These small motorized cars allowed trackside maintenance to be done quickly, easily, and cheaply. It was just a matter of piling on the needed equipment and men and puttering to wherever they needed to be. The workmen are (*left to right*) Fortino Rodriquez, Santiago Fresquez, Santiago Pacheco, Adam Martinez, and Manuel Rodriquez. (*Friends, Dorman Collection, RD089-132*)

TRACKING THE CHILI LINE TODAY

Driving south on NM68 from Embudo, the Chili Line grade is very visible on the west bank of the Rio Grande, directly below the power poles. As the highway enters Velarde, there is no sign of the Chili Line bridge that once crossed the river here into the village. The bridge was already in great disrepair when Robert W. Richardson photographed it on July 2, 1941. However, there is still evidence of why the Chili Line entered Velarde with several fruit and produce stands in town, and a large warehouse on the east side of the highway. The Chili Line hauled the bounty of the Velarde valley out to Santa Fe and Denver.

Traveling south of Velarde on NM68, the driver is faced with a choice of following the highway on to Española or tracking the Chili Line on the west bank of the Rio Grande. There is no evidence of the Chili Line grade remaining on the west bank, but the road there does approximately follow the route of the line. To follow it, turn west (right) on Rio Arriba County Road 0057 about half a mile south of the Velarde Elementary School. The road heads west and crosses a bridge over the Rio Grande. Then turn left on Rio Arriba County Road 0059, where there is a sign indicating the location of Lyden. Drive south on Rd. 0059 for about 6 miles. This is where the Chili Line followed along the

Today, several fruit and vegetable stands still operate in Velarde from Memorial Day until the end of October. The Chili Line ran a spur into Velarde to pick up the produce grown in the valley. (*Photograph by the author*)

west bank of the Rio Grande. The vegetation along the riverbank is quite dense today, and there are few views of the river, but there is one extensive view just past milepost 4.

Continue south on Rd. 0059 until its junction with NM74. A left turn will take you back across the Rio Grande and into San Juan Pueblo (today known by its native name Ohkay Owingeh). A right turn will take you into the village of Chamita. Where this road intersection is today, the Chili Line continued directly south, crossing the Rio Chama in about 1 mile. The Chili Line then stayed on the west bank of the Rio Grande into Española, about 5 miles farther south. To view the plaza of San Juan Pueblo, turn right off NM74 immediately past the elementary school where the road goes up a steep hill. A stop at the pueblo visitor center is well worth the time. Then return to NM74 and follow it to the intersection with NM68. Turn right and take NM68 into Española.

In Española, stay on NM 68 until its intersection with US84/285. Turn right here, and you will be on Paseo de Oñate, which was the original main street of Española. The highway soon crosses the Rio Grande and intersects with NM30. It is at this intersection where the Chili Line depot stood, with a spur to the Bond Company store. In 1890, there was also a roundhouse, windmill, water tank, a bunk house, a section house with a kitchen, a tool house, stock pens, and two warehouses. The siding could accommodate

Tracking the Chili Line Railroad to Santa Fe

Rio Arriba County Road 0059 runs along the Rio Grande between Lyden and Chamita. This was the grade for the Chili Line. (*Photograph by the author*)

The abandoned Railroad Ave. Grill still stands on Railroad Avenue in Española. It used a pictorial image of the Chili Line on its sign. The railroad ran directly in front of this building. (*Photograph by the author*)

thirty-seven cars. All of this is now gone. There is still some evidence of the Chili Line, however. At the intersection with NM30, take an immediate right turn to Railroad Avenue. The Chili Line ran here. The only evidence left is the name of the street and an old abandoned cafe, the Railroad Ave. Grill. To continue tracking the Chili Line, turn back to the intersection of US84/285 and NM30. Take NM30 south to Santa Clara Pueblo. This will be covered in the next chapter.

COLORFUL CHARACTERS OF THE CHILI LINE: THE BOND FAMILY OF ESPAÑOLA

In 1883, a couple of brothers from Quebec, Canada, arrived in Española seeking their fortune in the wild west. They were George W. and Frank Bond. The new railroad had attracted their attention, as well as the large sheep and cattle grazing areas near Española. They wanted to get into the mercantile business, so they purchased the established business of Scott and Whitehead. Then they purchased 40 acres of land adjacent to the railroad station and built facilities for shipping livestock there. Thus the G.W. Bond & Bro. mercantile business was established.

The brothers were extremely successful, and they built the largest business and the largest house in Española. They gradually transformed the local economy from trading to cash. The Bonds brought in manufactured goods such as sewing machines, kitchen ovens, and farming implements on the Chili Line, and they shipped out sheep, cattle, wool, and all kinds of fruits and vegetables produced in the valley. They shipped 200 carloads of apples per year, and "during the 1920s the Bond Brothers out of Española consigned and shipped 200,000 pounds of piñon nuts to San Francisco."[10] They paid the local farmers and ranchers cash for their products, then received cash in return as the farmers and ranchers purchased manufactured goods in the Bond store. The Bonds virtually controlled all shipping on the Chili Line in to and out of Española.

Frank Bond married May Anna Caffall from Pueblo, Colorado, in 1887, and they had three children—Maude, Amy, and Richard Franklin. Also in that year, Frank built the first two adobe rooms of the Bond House. As the family and the business grew, Frank added a major two-story addition to the house in 1911, which created a nine-room, 6,000-sq. foot residence, the most dominant in the valley. This is the Bond House, which still stands today at 706 Bond Street, just up the hill from where the train station and the Bond store were located.

When Frank Bond died in 1945, the family fortune went to his son Richard Franklin and his wife Ethel Moulton. Richard Franklin died in 1954, and Ethel sold the house to the City of Española in 1957 for $10,000. The house was used for city offices until the new City Hall was built in 1978. The Bond House was added to the National Register of Historic Places in 1980. It then became the Bond House Museum and Cultural Center. The San Gabriel Historical Society opened its first exhibit there on March 7, 1982. The Bond House was restored to its original condition in 2000, funded by a grant of $1.4 million. The Bond House Museum is open Monday–Friday afternoons, and it has excellent displays of Chili Line material and photographs of early Española.

Most lucrative of all for the Bond mercantile was the sheep business. They purchased sheep and secured grazing rights in Valles Caldera, above what is now Los Alamos.

The Bond House Museum sits on a hill overlooking Española and the Sangre de Cristo Mountains to the east. Frank Bond built the first two adobe rooms on the left in 1887 and added the two-story structure on the right in 1911. (*Photograph by the author*)

Their sheep herd consisted of as many as 32,000 animals. Valles Caldera is a high mountain pasture in the crater of an extinct volcano. Some 100,000 acres of this area was part of the Baca land grant from the Mexican Government to the Baca family, later affirmed by the United States Government.

In 1917, George and Frank Bond leased land on the Baca grant to graze their sheep. In 1925, the Bonds purchased this land for $500,000. The Bond family owned this land until the end of 1962. On January 1, 1963, the family sold it to James Dunigan, a Texas oilman, for $2.3 million. His sons sold 89,000 acres of the Baca Ranch to the U.S. Forest Service in 1999 for the incredible sum of $101 million. This land became the Valles Caldera National Preserve in 2000, managed by the Valles Caldera Trust established by the Forest Service. In 2014, management was transferred to the National Park Service. Author William DeBuys noted:

> The federal acquisition of the Baca was an extraordinary achievement. It required willing sellers, tireless and top-flight agency work, bipartisan political commitment, and a colossal sum of money … It exceeds the amount the Forest Service had paid for any other single acquisition in its long history.[11]

Thus, we have the Bond family to thank for originally acquiring this land on the Baca Ranch, which forms the beautiful national park located there now.

6

ESPAÑOLA TO BUCKMAN

With Chili Line construction halted at Española in 1881 due to the Treaty of Boston, rail passengers had to debark there and take the stagecoach line to Santa Fe. This inconvenience continued for six years until the Texas, Santa Fe and Northern Railroad reached Española from Santa Fe on January 8, 1887. Passengers then simply had to transfer from one train to another. The trackage to Santa Fe officially became part of the D&RG system on August 1, 1908.

In Española, Chili Line passengers probably encountered Santa Claran Pueblo natives selling their pottery. Santa Clara was just 3 miles south of Española on the Chili Line.

> Santa Clara ancestors most likely migrated south from the Four Corners region to the Chama River Valley, eventually reaching the Pajarito Plateau on the eastern slope of the Jemez Mountains between A.D. 1100 and 1300.[1]

There they built cliff dwellings at a site called Puye. The Puye Cliff Dwelling ruins can be visited by travelers today. In the 1500s, a major drought forced the cliff dwellers off the plateau and down to the west bank of the Rio Grande. Here, Santa Clara Pueblo was established, furnished with abundant water from the Rio Grande for growing crops.

Traveling south on the Chili Line, passengers next would view San Ildefonso Pueblo some 4 miles from Santa Clara. However, San Ildefonso is on the east bank of the Rio Grande, so this would have been a much more distant view. On the west bank, the D&RG had a siding that could accommodate sixteen cars on its 661 feet length. Like Santa Clara, the ancestors of San Ildefonso arrived on the Pajarito Plateau around A.D. 1100–1300. There, they built several villages, and the ruins of one, Tsankawi, can be visited today by hiking a trail in Bandelier National Monument. Also, like Santa Clara, drought forced the San Ildefonso ancestors off the plateau and down to the Rio Grande. San Ildefonso is famous for its pottery, and the black pottery of Maria Martinez is well-known worldwide.

Some 2 miles south of San Ildefonso, the Chili Line reached the tiny hamlet of Otowi. "The name of this site on the Rio Grande comes from a Tewa word meaning 'gap where water sinks,' referring to a place where the water of nearby Pueblo Creek often sinks into the sand."[2] The Chili Line had a freight car body depot at Otowi, a water tank, and a

Tracking the Chili Line Railroad to Santa Fe

Santa Clara Indians selling pottery to a Chili Line work crew in Española. Santa Clara Pueblo is 3 miles south of Española. (*Friends, Dorman Collection, RD088-005*)

With the Chili Line track in the foreground, Santa Clara Pueblo can be seen between the track and the Rio Grande in the distance. Photograph is by Edward S. Curtis in 1905. (*Friends, Dorman Collection, RDS071-184*)

siding of 215 feet, which could accommodate eight cars. The Chili Line crossed the Rio Grande at Otowi on a bridge built by the Texas, Santa Fe and Northern Railroad in 1886. This marked the first time that the Chili Line crossed to the east bank of the Rio Grande, and it remained on the east bank through White Rock Canyon to Buckman, where it then left the riverbank and ascended to Santa Fe. In 1921, a one-lane wooden suspension bridge was constructed across the Rio Grande just north of the railroad bridge. It carried a primitive road up to the Pajarito Plateau and Bandelier National Monument.

When the Chili Line was abandoned in 1941, "Everyone missed the line," wrote Oliver La Farge:

> The whistle of the northbound train as it came around the cliffs near Otowi was the signal for San Ildefonso and Santa Clara Indians and the Spanish-Americans nearby to leave their fields and go to lunch. As one Indian told me, once the train stopped running, they would either have to buy watches or go back to the ways of their ancestors and plant sticks in the ground to tell them when it was noon.[3]

Otowi was the site where freight supplies from Santa Fe were dropped off by the Chili Line for the Los Alamos Boys' Ranch School, located about 10 miles west of there on the plateau. The Ranch School was established in 1917 by Ashley Pond as an exclusive school for boys where they could get a college prep education and experience the outdoors through hiking, camping, and hunting. The boys were encouraged to bring a horse, a rifle, and a dog with them to the school.

> When Ashley Pond … made his 1917 agreement with the D & RG for the establishment of a railroad stop at the Otowi crossing, his freight and mail station consisted of a converted boxcar station house and an eight-car siding.[4]

When the railroad stop was established, Pond hired "Shorty" Pelazu to manage the station and unload the freight, storing it there until a truck came down from the school (usually three times a week) to pick it up. Pelazu, a former lumberman from farther south in the canyon, moved his two-room frame home to land he rented at Otowi from Julian and Maria Martinez of San Ildefonso.

After the building of the 1921 one-lane wooden highway bridge over the Rio Grande, Shorty increased his income by using part of his home as a general store with a gasoline pump outside for traveling motorists. Shorty disappeared in the late 1920s, probably due to his bootlegging activities. Adam Martinez (son of Julian and Maria) was then hired to be the station master. Adam later became homesick for his native San Ildefonso Pueblo, and he moved his family back there. This left the freight unguarded at the station. Desperate to find a new employee to guard the station, A. J. Connell, director of the Los Alamos Boys' Ranch School, went to Santa Fe. There he met Edith Warner at the La Fonda Hotel, and he hired her on the spot despite her diminutive size. Edith moved to Otowi on May 1, 1928. She settled into Shorty's former frame home at Otowi and eventually opened a tearoom there to serve travelers on their way to Bandelier National Monument, and later the scientists on their way to the new Los Alamos atomic labs. Edith is profiled below as a "colorful character" of the Chili Line.

Just south of Otowi, the Chili Line entered the spectacular White Rock Canyon, bursting through at Buckman, 4 miles away. Buckman was named for H. F. Buckman,

C-16 no. 207 stops at the Otowi boxcar depot to unload freight, *c.* 1920s. (*Friends, Dorman Collection, RDS077-037*)

The northbound Chili Line train is stopped at Otowi to unload freight. The date has to be 1921 or later, as the highway bridge behind the train was constructed in 1921. The truck parked next to the train picks up supplies for the Los Alamos Boys' Ranch School. (*Friends, Dorman Collection, RD088-030*)

Freight agent Edith Warner stands on the steps of the Otowi depot in 1930. Miss Warner was hired by the Los Alamos Boys' Ranch School in 1928 to oversee unloading and storage of freight brought in by the Chili Line for the school. (*Friends, Dorman Collection, RD089-052*)

The southbound Chili Line train pulls into Otowi, c. 1920s. There is a truck parked next to the train, and a boxcar on the siding beyond the truck. *(Friends, Dorman Collection, RD088-035)*

This Robert W. Richardson photograph taken on July 2, 1941, shows the bridges at Otowi looking west across the Rio Grande. The Chili Line bridge was constructed in 1886 by the Texas, Santa Fe and Northern Railroad, and the highway bridge to the right was constructed in 1921. The Highway bridge still exists today (though no longer used), but the Chili bridge was dismantled late in 1941. (*Friends, Dorman Collection, RD088-038*)

This northbound Chili Line train is crossing the Rio Grande at Otowi, c. 1930. This mixed train had a consist of a boxcar, R.P.O., and passenger coach. (*Friends, Dorman Collection, RDS071-179*)

Stacks of lumber are piled at Buckman on the east side of the Rio Grande, c. 1910, awaiting shipment on the Chili Line. Timber was harvested west of the Rio Grande on the Pajarito Plateau, and brought down by wagons, crossing the Rio Grande on a primitive bridge to the rail line at Buckman. When the highway bridge was constructed at Otowi in 1921, the Buckman bridge was abandoned. (*Friends, Dorman Collection, RD088-014*)

Tracking the Chili Line Railroad to Santa Fe

These folks were gathered at the Buckman depot/post office on December 7, 1914, waiting for the train. (*Friends, Dorman Collection, RD089-049*)

an Oregon lumberman, who ran a timber cutting and sawmill operation on the Pajarito Plateau. The cut timber was hauled down to a primitive bridge across the Rio Grande to the railroad siding on the other side, where it was shipped out on the Chili Line. The D&RG siding was 751 feet long, with a capacity of sixteen cars. There were also stock pens at Buckman, as well as a small village served by a post office. At Buckman, the Chili Line left the Rio Grande and began its 21-mile climb uphill to Santa Fe.

TRACKING THE CHILI LINE TODAY

Heading south out of Española on NM30, the traveler will be following the route of the Chili Line along the west bank of the Rio Grande. In 3 miles, turn left into Santa Clara Pueblo for an opportunity to buy pottery and see the Santa Clara church, built of adobe in 1918. The Spanish named the pueblo and the church Santa Clara:

> For Saint Clare, founder of the Order of Saint Clares, an order of Franciscan cloistered nuns. According to legend, Saint Clare made clothing for Saint Francis of Assisi, and thus she is the patron saint of weavers and seamstresses.[5]

Return to NM30 and continue south for about 1 mile, then turn right at the gas station if you are interested in visiting the Puye Cliff Dwellings, the ancient ancestral home of the Santa Clarans.

From Santa Clara, continue south on NM30, following the Rio Grande. There are some homes of the San Ildefonso Pueblo on the west bank, but the pueblo itself and the main village are on the east side of the river, not visible from the highway. Some 6 miles south of Santa Clara, you will reach the intersection of NM502. A left turn here will lead a quarter of a mile to the Rio Grande where you can view the old highway suspension bridge on the south side of the road, and Edith Warner's tearoom and guest house on the north side.

Continuing east on NM502, you will come upon the entrance road to San Ildefonso Pueblo on the left. It is well worth a visit to this historic pueblo, home of famed potter Maria Martinez. There are pottery shops here, along with the plaza with its circular kiva, and the spectacular San Ildefonso Church built in 1968 on the site of the original church. Return to NM 502 and continue east to the intersection with US84/285. Turn right here for the 16-mile journey south to Santa Fe.

The D&RG actually did some grading for the Chili Line in White Rock Canyon in anticipation of the line heading south from Española to Santa Fe. However, that work was stopped with the Treaty of Boston in 1880. The Texas, Santa Fe and Northern Railroad used this grade while building north from Santa Fe. Travelers today can view this grade by taking NM502 west from NM30 to the intersection with NM4. Follow NM4 into the town of White Rock, then look for signs there leading to Overlook Park. Continue following the signs through the neighborhoods to where the road ends at the parking lot in Overlook Park. From there, it is a very short walk to the overlook. Here, there is a spectacular view of White Rock Canyon far below with the Rio Grande heading south through the canyon. The grade of the Chili Line exiting White Rock Canyon can be seen on the horizon to the east following along the current road to Buckman.

To reach Buckman today, you will have to retrace your drive—NM4 to NM502 to US84/285 to Santa Fe. At the northern boundary of Santa Fe, take NM599 west to Camino La Tierra. Turn right at the exit here and stay on Camino La Tierra until its junction with Old Buckman Road in approximately 4 miles. Turn right and follow the graveled Old Buckman Road, eventually passing Diablo Canyon on the west. Continue on north to where the road ends at the Rio Grande. Here is where the old settlement of Buckman was located. Upon entering NM599, it is 17 miles to Diablo Canyon, and from there, it is another 3 miles to Buckman. At Buckman, just before the road dead ends at the river, there is a hiking trail leading off to the right (north). This trail follows the Chili Line grade for about 1.5 miles north to the border with San Ildefonso Pueblo, which is indicated by a fence, but there is no sign. Here, the hiker must turn around, as entry into San Ildefonso Pueblo is prohibited.

At Buckman where the road dead ends at the Rio Grande, travelers today can hike the Chili Line grade for 1.5 miles on trail 26X. The peak outlined in the distance matches exactly with the peak shown in the historic photograph on page 95 of lumber piled at Buckman. (*Photograph by the author*)

Opposite above: A snowy day on January 5, 2019, shows the old Otowi highway suspension bridge still standing over a frozen Rio Grande below. The view is to the east. (*Photograph by the author*)

Opposite below: This January 5, 2019 view shows the remains of buildings in Edith Warner's complex at Otowi where she had a home, a store, a tearoom, and a guest house. (*Photograph by the author*)

Tracking the Chili Line Railroad to Santa Fe

This April 2019 view looks east from Overlook Park at the town of White Rock, high above the Rio Grande. In the upper midsection the straight line from right to left is the present road to Buckman, which is just off the photograph to the left. The Chili Line followed this same grade up to Santa Fe from Buckman's site on the river. The Sangre de Cristo Mountains rise majestically on the upper horizon with the snow-covered slopes of Ski Santa Fe visible on the peak to the right. (*Photograph by the author*)

COLORFUL CHARACTERS OF THE CHILI LINE: EDITH WARNER

The story of Edith Warner at Otowi is remarkable. It would be unbelievable, except it is true. That she survived and actually thrived there is testament to her sheer determination, unwillingness to be defeated, and her love for the land and the San Ildefonso natives.

Edith was born in 1882 in Pennsylvania, the daughter of a Presbyterian minister. She became a schoolteacher and then a secretary. She was not satisfied with either career. Due to health issues, she headed for the dry climate of New Mexico in 1922. She met John and Martha Boyd and stayed with them at their ranch in Frijoles Canyon (Bandelier National Monument). The Boyds introduced Edith to the people and geography of the area, which she fell in love with. She wanted to stay in the area but was unable to find a job, so she was forced to move back east. Her health deteriorated again, and she headed west to Denver and eventually Santa Fe. Unable to find work, she was down to her last funds when she met A. J. Connell of the Los Alamos Boys' Ranch School in the lobby of the La Fonda Hotel. When he offered her the job of freight agent at Otowi in 1928, she readily accepted even though the salary was only $25 a month. From that amount, she was expected to pay rent to Julian and Maria Martinez for the house at Otowi, and pay the wages of Adam Martinez, their son, who would come over from San Ildefonso Pueblo to help unload the freight from the Chili Line. Edith, wondering how she would have any money for herself, was told by Connell that she would be entitled to the profits from the general store and gas pump at Otowi.

Edith Warner poses with a truck at her gasoline pump and general store at Otowi in 1930. (*Friends, Dorman Collection, RD088-027*)

Edith Warner moved into the frame house at Otowi on May 1, 1928. She found it to be extremely dilapidated and filthy. She immediately began cleaning the place, enlisting Adam Martinez to help with repairs on the home in addition to his freight duties. Edith soon found that her meager earnings from the general store were not enough to survive upon. She was in love with New Mexico and the San Ildefonso Indians and was desperate not to be forced to leave again. She came up with the idea of establishing a tearoom in the home to serve passing travelers and bring in more income. "With the aid of Adam Martinez and his great-uncle Atilano Montoya, she was able to remodel the middle room into a 'tearoom' where she could serve meals to her guests."[6] The tea room was a great success, and it enabled Edith to stay at Otowi until her death in 1951.

Atilano (Tilano) Montoya was nearly sixty years old when he came to live with Edith Warner. His room was at the back of the house, and he was her constant companion until her death. He served as handyman and handled everyday chores such as gathering wood for the stove. In 1934, he helped Edith build a three-room adobe guest house. There were two bedrooms with a living room in the middle with a fireplace. Edith wanted to provide a "retreat" for travelers so that they could enjoy the wonderful climate and scenery of New Mexico. It also provided her with an extra source of income.

In December 1942, the U.S. Army evicted the Los Alamos Boys' Ranch School from the Pajarito Plateau to develop a top secret laboratory for the development of an atomic bomb. J. Robert Oppenheimer, the director of this "Manhattan Project" to develop the atomic bomb, recalled how he first came to the area and how he met Edith:

> I first knew the Pajarito Plateau in the summer of 1922, when we took a pack trip up from Frijoles and into Valle Grande. We came back to it often from our ranch in the Pecos. In the summer of 1937 I first stopped at Edith Warner's tea room. I was on a pack trip with my brother and sister-in-law … We had tea and chocolate cake and talk; it was my first unforgettable meeting. I remember that in the summer of 1941 I brought my wife over to introduce her to Edith. By early 1943 we came to Los Alamos, and very early we stopped to talk to her … We saw her regularly after that … We were very fond of her.[7]

Though the Chili Line had been abandoned in 1941, the influx of hundreds of scientists to Los Alamos in 1943 brought increasing business to Edith Warner's tearoom. She and Tilano decided to add an adobe dining room to their frame house in order to accommodate the newcomers.

> Though the project to construct the atomic bomb was carried out in the strictest secrecy, the necessity of allowing his fellow scientists to leave their isolated surroundings was recognized by Director J. Robert Oppenheimer, who occasionally permitted small groups to leave "The Hill" and eat at the "tearoom." Edith Warner had known Oppenheimer from earlier days when the young scientist had come over from his ranch in the Pecos River Valley to have supper in her tearoom. However, she never learned the real names of her brilliant and charming guests until the end of the war, and it was only after the destruction of Hiroshima and Nagasaki that that she was able to write her friends telling them that such men as Bohr, Conant, Oppenheimer, Compton, Fermi, Allison, Teller and Parsons had spent long hours talking and eating in her tearoom.[8]

After the war, Edith and Tilano hoped for a bit more seclusion and quiet in their elder years. However, this was not to be, since in 1947, a steel highway bridge was built to replace the old one-lane bridge across the Rio Grande. The new highway would cut right across Edith's front yard. Deciding that they could not stand the noise, Edith and Tilano selected a site for a new home half a mile farther west. Edith's friends from Los Alamos and San Ildefonso built the adobe home for her. Tilano managed the project and had his hands full, as the scientists and the Indians could not always agree about how to get things done. The Indians had hundreds of years more experience building adobe homes though, and their suggestions usually prevailed. Friends continued to provide for the couple in the new home as the tearoom was permanently closed.

In 1951, Edith became gravely ill with cancer. She died on May 4, 1951, in the place that she had loved for so long. "This shy little spinster from Pennsylvania lived for more than twenty years as neighbor to the Indians of San Ildefonso Pueblo, and when she died they buried her among them."[9] Tilano, then in his eighties, passed away two years later. The remains of the tearoom and guest house can still be seen along the northwest bank of the Rio Grande at Otowi today. Those remains and the old one-lane highway bridge, which still stands, were added to the National Register of Historic Places in 1975.

7

BUCKMAN TO SANTA FE

The Chili Line stopped in Buckman to pick up lumber and cattle. In 1921, the plank road bridge across the Rio Grande at Buckman collapsed, and the village of Buckman soon became a ghost town as it was cut off from the timber operations on the Pajarito Plateau. There is no sign of it today, but there is a large new operation at Buckman called the Buckman Direct Diversion Project. The project was completed in 2010 by the City and County of Santa Fe to divert water from the Rio Grande, treat the water, and pump it uphill to the city. Travelers to Buckman today will see a number of wells and pumping stations along the drive. Where the road ends at the Rio Grande, there is a very large structure which is the diversion facility.

Some 3 miles south of Buckman, the Chili Line passed Diablo Canyon off to the west. Famed western photographer William Henry Jackson photographed the Chili Line passing Diablo Canyon. Though there is no exact date on Jackson's photograph, he did identify it as "Canon Diablo, Santa Fe Southern Railway." Since the Santa Fe Southern did not purchase the Texas, Santa Fe and Northern Railroad until 1889, the photograph must have been taken in 1889 or 1890.

Some 10 miles from Buckman, the Chili Line had a siding at Jacona, which was 958 feet long and could accommodate twenty-three cars. This siding served as a pick-up and drop-off point for cattle cars from the ranching operations in the area. One of the largest ranches was the Buckman Ranch, which consisted of 31,000 acres. It was purchased from the Española Bond family in 1960 by Suzanne Hoyt and her husband, Bob Weil. They developed parts of the ranch into what are now the subdivisions of Las Campanas and La Tierra, northwest of Santa Fe.

There is an elevation gain of a little over 1,500 feet from Buckman to Santa Fe. From Jacona siding to Santa Fe, the Chili Line ascended the final 12 miles. Author Forest Crossen rode the Chili Line to Santa Fe in 1941, and described this view of the city as it came into view:

> On a gentle rise of land below us she lay, her adobe walls soft in the late afternoon sun, alone and aloof, yet strangely warm, like a queen drawn apart from the mad clash of our lives, Santa Fe, golden city.[1]

Photographer William Henry Jackson had a special excursion train on the Chili Line in 1889 or 1890. He photographed the train on a trestle over an arroyo just east of the entrance to Diablo Canyon. (*History Colorado, 86.200.618*)

K-28 no. 475 crosses a small arroyo in the vicinity of Buckman/Diablo Canyon, c. 1930. The arroyos ran intermittently with water after large rainstorms that were common in summer, necessitating the use of many trestles. Between Buckman and Jacona, there were twenty-eight trestles, and between Jacona and Santa Fe there were nineteen trestles. (*Friends, Dorman Collection, RD089-011*)

Author John Gjevre also noted the allure of Santa Fe:

> From the earliest Chili Line days, Santa Fe was an artist's goal, where high fresh mountain air, the triple cultures of the Pueblo Indians, the Spanish Americans and the Anglos could dwell, commingle and laugh at one another's foibles.[2]

With the founding of the Taos Society of Artists in 1915, and the opening of the New Mexico Museum of Art in Santa Fe in 1917, artists were indeed discovering the allure of this beautiful place. However, not all early travelers viewed Santa Fe with an "artist's eye." Englishman Frederic Ruxton visited Santa Fe in 1846, and had a distinctly different view:

> Santa Fe … is a wretched collection … of mud houses … the appearance of the town defies description, and I can compare it to nothing but a dilapidated brick-kiln or a prairie dog town. The inhabitants are worthy of their city, and a more miserable, vicious-looking population it would be impossible to imagine … Although I (had) determined to remain some time in Santa Fe … I was so disgusted with the filth of the town, and the disreputable society a stranger was forced into, that in a very few days I once more packed my mules and proceeded to the north.[3]

Interestingly, Santa Fe in some ways today still resembles a "collection of mud houses." However, there are few actual adobe homes today. The building technique favored across the city is to make the homes look like adobe homes by constructing the houses with a lumber frame, composition board walls, flat roofs, and then plastering the outer walls with stucco. Most of the stucco applied is a shade of brown, so to some, the town may still look like a "brick-kiln."

The Chili Line came into Santa Fe on Jefferson Street, now called Guadalupe Street. Before crossing the Santa Fe River (actually no more than a creek about 3 feet wide when there is any water in it at all), the D&RG constructed a station and a three-stall engine house just north of the river. The passenger depot remained at that spot until a new brick depot was constructed in 1903 about two-thirds of a mile south, on the west side of Guadalupe Street. This was the "Union Depot," so called because the D&RG shared the depot with the Santa Fe Central Railroad. This was a standard gauge railroad, so the Union Depot had narrow-gauge tracks for the Chili Line coming in from the north, and standard-gauge tracks heading south out of the depot for the Santa Fe Central. The Santa Fe Central Railroad became the New Mexico Central Railroad in 1908 when the Santa Fe Central joined the Albuquerque Eastern Railway. Also in 1908, the Denver and Rio Grande Railroad merged with the Rio Grande Western (the Utah extension) to become the Denver and Rio Grande Western Railway.

When the Chili Line crossed the Santa Fe River in 1903, it immediately passed the Santuario de Guadalupe on the west side. The Chili Line then continued right down the center of Guadalupe Street to the new depot. Just to the west of that depot stood the Atchison, Topeka, and Santa Fe Railroad depot. The AT&SF had arrived in Santa Fe on February 9, 1880, "to be properly welcomed by flowing speeches and a parade."[4] Platforms were constructed between the two depots so that freight could be unloaded from the narrow gauge Chili Line and transferred across a platform to be loaded on the standard-gauge AT&SF for shipment to points east and west.

Above: Passing in front of the Santuario de Guadalupe in Santa Fe, this northbound Chili Line train is crossing the bridge over the Santa Fe River, *c.* 1941. (*Friends, Dorman Collection, RDS071-160*)

Right: Chili Line Railroad Crossing sign directly in front of the Santuario de Guadalupe. After crossing the Santa Fe River heading south, the Chili Line ran directly down the middle of Guadalupe Street. Photograph by F. D. Nichols for the Historic American Buildings Survey, 1934. (*Library of Congress, HABS NM, 25-SANFE,8*)

Robert W. Richardson photographed this Chili Line train on Guadalupe Street on July 3, 1941. (*Friends, Dorman Collection, RD089-092*)

K-28 no. 473 has just arrived at the Santa Fe depot with an R.P.O. car and a passenger coach, c. 1940. (*Friends, Dorman Collection, RD089-058*)

In this photograph by Richard B. Jackson, the mixed Chili Line southbound train is arriving at the Santa Fe depot with a Railway Express truck awaiting its arrival. (*Friends, Dorman Collection, RD089-139*)

Some years after the abandonment of the Chili Line in 1941, the vacant depot sits forlornly on Guadalupe Street. The Atchison Topeka and Santa Fe depot is the white building to the left. A view like this of both depots is not possible today because Tomasita's Restaurant (which occupies the Chili Line depot) has added on to the building extensively at the back (where the sewer pipe is), blocking the view. (*Friends, Dorman Collection, RDS071-043*)

TRACKING THE CHILI LINE TODAY

In the previous chapter, we described the route to reach Buckman today. Now, retracing the same route back south to Santa Fe, the towering cliffs of Diablo Canyon appear on the right (west) about 3 miles from Buckman. There is a parking area and primitive campground at Diablo Canyon, with a hiking trail leading from there three miles back to the Rio Grande. The entrance to Diablo Canyon appears much the same as it did in William Henry Jackson's 1889 photograph. Continuing on south from Diablo Canyon, the Chili Line grade is quite visible on the left-hand side. At points along the way, there are still wooden pilings for the Chili Line bridge trestles over the arroyos.

Continue following Old Buckman Road back to Camino La Tierra and on to NM599. The route passes through the old Buckman Ranch, most of which has now been transformed into the subdivisions of La Tierra and Las Campanas. Take a left turn at NM599 back to the intersection with US84/285 and turn right heading downhill to Santa Fe. At the upcoming intersection with St. Francis Street, take the left exit to Guadalupe Street.

The Chili Line followed Guadalupe Street across the Santa Fe River, passing by the Santuario de Guadalupe (Guadalupe Church) and on to the Union Depot a bit farther south. You can drive this same route today, but Guadalupe Street is quite busy with cars and pedestrians, so careful driving is necessary. The Chili Line depot still exists today and is inhabited by Tomasita's Restaurant. Parking there is restricted to restaurant customers, so to obtain photographs of the depot, either eat a meal at the restaurant or find a parking place at a meter on the street.

Just to the west of Tomasita's is the New Mexico Railrunner terminus at the old AT&SF depot. The Railrunner provides commuter service from Santa Fe to Albuquerque. Just to the south of the Chili Line depot is the large Gross, Kelly and Company Warehouse, which was built in 1913. It was designed by architect Isaac Hamilton Rapp, who helped develop the Spanish Pueblo Revival Style in New Mexico. He continued to display this style in the New Mexico Art Museum just off the Plaza, and the La Fonda Hotel east of the Plaza on San Francisco Street. The warehouse played an important role in Santa Fe's economy in the 1900s, providing storage space for manufactured products brought in by the Chili Line and the AT&SF and shipping out local products such as wool, hides, grain, potatoes, beans, and chili.

> The Company served the important function of expediting the transfer of eastern manufactured goods for the raw materials of the frontier. In an economy where cash was scarce, payment was often accepted in goods, such as livestock, wool, or hides that were, in turn, sold to eastern manufacturers … In this way the company provided a dependable market for local products.[5]

The Gross, Kelly and Company Warehouse today is the headquarters for one of Santa Fe's realty firms. In the past ten years, the city of Santa Fe has transformed the dilapidated railyard area into a delightful area of art galleries and restaurants, retail shops, and a movie theatre. Farmer's markets and arts and crafts fairs are held here on many weekends.

This contemporary view of Diablo Canyon matches almost perfectly with William Henry Jackson's 1889 photograph on page 105. One can almost visualize a Chili Line train on the gravel road in the foreground. (*Photograph by the author*)

Along Old Buckman Road, pilings for a Chili Line trestle crossing an arroyo can still be seen. Diablo Canyon is in the middle background. (*Photograph by the author*)

The Chili Line is long gone from Guadalupe Street and the bridge over the Santa Fe River, but the Santuario de Guadalupe still stands in the background in this contemporary view. (*Photograph by the author*)

The Chili Line depot at Guadalupe Street and Manhattan Avenue is now the home of Tomasita's Restaurant. (*Photograph by the author*)

COLORFUL CHARACTERS OF THE CHILI LINE: SANTA FE

Rather than one particular human character, we will spotlight the city of Santa Fe itself as a character with its many historic buildings, one of which dates back to the founding date of the city in 1610. We can only give a glimpse of Santa Fe's long, complicated history here, with the hope that readers will be inspired to explore further on their own.

Santa Fe was founded in the winter of 1609–1610 when Don Pedro de Peralta (the third Spanish governor of the province of New Mexico) settled here. The first Spanish "capital" founded by Don Juan de Oñate at San Gabriel (near Chamita) was abandoned, and the settlers migrated here to the banks of the Santa Fe River. In 1610, Peralta built the Palace of the Governors as a home and fortress for the Spanish governors, and thus this became the official capital of New Mexico.

The Spanish were rousted from New Mexico in 1680 by the Pueblo Revolt. This was an uprising of all the New Mexico Pueblos against their harsh Spanish masters. The revolt was so successful that it drove the Spanish completely out of New Mexico and south to Old Mexico for a period of twelve years. The Puebloans took over the Palace of the Governors and established homes and a native worship center there. The Spanish, led by Don Diego de Vargas, successfully recaptured New Mexico from the Indians in 1692. Santa Fe then continued to serve as the seat of government for the Spanish governors, until 1821, when Mexico became independent from Spain.

It then served Mexico's New Mexican governors until 1848, when New Mexico became part of the United States as a result of Mexico's loss of a war with the United States. From 1848, the Palace of the Governors was occupied by American Territorial governors "until 1907, when the present executive mansion was built … By act of the New Mexico legislature in 1909, the Museum of New Mexico was established and located in the Palace."[6] New Mexico became a state in 1912, and Santa Fe officially became the state capital.

The Palace of the Governors still stands and is today part of the New Mexico History Museum. It occasionally undergoes restoration, with a 2019 restoration to install modern climate control systems to protect the historic artifacts housed there. The Palace of the Governors is located on the northern portion of the downtown Santa Fe Plaza. Native Americans can be found every day selling their handmade crafts on the sidewalk in front of the Palace. They must be granted a license by the History Museum to sell their wares.

One of the first historic buildings to greet Chili Line passengers traveling along Guadalupe Street was the Santuario de Guadalupe, which was seen on the right side of the train as it headed south to the depot. It was built as a shrine to honor the Mexican Virgin, Our Lady of Guadalupe (Nuestra Señora de Guadalupe), and the license to build a chapel on this spot was granted by the archdiocese on October 14, 1795. Being constructed at this later date, the church escaped the desecration of the Pueblo Revolt of 1680, which had destroyed almost every other Spanish church in New Mexico. The church has undergone many transformations over the years, and the present Santuario, was reconstructed in the California mission style after a fire in 1922, with a new pitched roof, a bell tower, and a wooden floor. This is the style seen in most Chili Line photographs of the church, and it remains today. Because of a great increase in the number of parishioners, a massive new church was completed behind the Santuario in December 1961. The Santuario is now mainly a museum open to the public, with special masses at certain times.

Tracking the Chili Line Railroad to Santa Fe

A front view of the Palace of the Governors taken by M. James Slack for the Historic American Buildings Survey on March 21, 1934. Cars are no longer allowed to park in front of the building as the street is closed to vehicular traffic. (*Library of Congress, HABS NM 25-SANFE, 2*)

A contemporary view of the Santuario de Guadalupe. The Chili Line used to run right in front of it. (*Photograph by the author*)

Buckman to Santa Fe

Chili Line passengers came from Denver, southern Colorado, and Northern New Mexico. Many of them took time to visit the historic sites in Santa Fe after their arrival, including Indians from the Northern New Mexico pueblos.

> Indians were allowed to ride free, a "kind service" provided to somewhat compensate for the tracks bisecting their reservations. Indian families would board by waving down the train as it passed through the reservation, and would ride to either Santa Fe or Antonito, then turn around and come back. It was just good entertainment![7]

Besides the Palace of the Governors and the Santuario de Guadalupe, visitors would often seek out the "oldest church in America," which was the San Miguel Chapel. Still standing, the chapel is just south of the Santa Fe River at the intersection of Old Santa Fe Trail and DeVargas Street. The earliest documentation of the existence of the chapel was 1628, but "oral history holds that San Miguel Chapel was built around 1610, and it has been rebuilt and restored several times over the past 400 years."[8]

Most of the chapel was destroyed in the Pueblo Revolt of 1680, but upon the Spanish re-conquest, Governor General Don Diego de Vargas ordered the repair and restoration of the chapel. Work was completed in 1710, with major repairs undertaken again in 1798. In 1859, Archbishop Jean Baptiste Lamy of Santa Fe purchased the chapel and adjacent land where he hoped to build a Christian Brothers school. The school was built, but the Brothers could not afford the upkeep of the chapel, and it was threatened with demolition in 1887. Community leaders then stepped in to save the chapel, and

This *c.* 1880 view of the San Miguel Chapel shows repair work being done on the front entrance. The adobe bricks are seen very clearly in this photo. (*Library of Congress, LC-DIG-PPMSCA-39914*)

the stone buttresses were added at this time to shore up the adobe walls of the chapel. Another major restoration was completed in 1955. San Miguel Chapel remains one of the best examples of preserved adobe architecture in Santa Fe today.

Saint Francis Cathedral is located one block south and one block east of the Santa Fe Plaza on San Francisco Street. Upon first viewing this church, a visitor might conclude that a European cathedral had been plunked in the middle of Santa Fe. The visitor would not be far from wrong because the cathedral was built beginning in 1869 by Archbishop Jean Baptiste Lamy who was from France.

> Lamy began construction of the stone St. Francis Cathedral in 1869 on the approximate site of earlier churches. Designed in the Romanesque style of Bishop Lamy's native Auvergne, the main building is architecturally foreign to Santa Fe's Spanish heritage and Indian background.[9]

Marie Romero Cash noted that "This Romanesque style was viewed by Lamy as a necessary modernization which would relieve Santa Fe of its crude pueblo look and give it badly needed sophistication."[10]

St. Francis Cathedral stands at the east end of San Francisco Street. This June 4, 2009, photo of the cathedral was taken by John Fowler of Placitas, NM. (*Courtesy of the photographer*)

The stone for the cathedral came from local quarries. The two towers were originally designed to have a 160-foot steeple on each tower, but these were never completed, probably due to a lack of funds. While the cathedral does not match Santa Fe's Spanish and Indian architecture, it is so grand and imposing that it remains one of the most visited sites in Santa Fe today.

EPILOGUE

It seems like the Chili Line always had problems during its sixty-one-year existence. Arroyos flooded and trestles were washed away. A fire burned down a trestle. There were frequent derailments, sometimes with disastrous consequences. "No part of the D & RG had less day-to-day maintenance than the Santa Fe Branch, and it really showed."[1]

By the mid-1930s, first the Great Depression, and then improved roads, automobiles, and trucks, all had taken a serious toll on the Chili Line. It was simply cheaper and faster to ship products by truck than on the Chili Line which poked along at about 17 miles per hour.

> There was a net loss from (Chili Line) branch operations of $243,000 during the five years prior to 1940. Passenger traffic had all but dried up with the line averaging but ten passengers per trip.[2]

The Interstate Commerce Commission ordered the abandonment of the Chili Line in 1941.

> During the summer of 1941 a fair amount of passenger service traffic was generated with the knowledge that the line was on its way to oblivion. Several all day parties were organized in Santa Fe—a ride to Embudo and return.[3]

The final run of the Chili Line was from Santa Fe to Antonito on September 1, 1941. Just about as soon as the train reached Antonito, the scrappers arrived, picking up the rails. Some locomotives were transferred to the D&RG's Antonito to Durango run, but several were sold to the U.S. Army. Had anyone envisioned in 1941 what would happen just two years later with the development of the atomic bomb labs at Los Alamos, perhaps the Chili Line could have been saved and used to transport the tremendous amount of materials needed to construct Los Alamos. Ah, hindsight- the bane of history.

Perhaps we can best give a final salute to the Chili Line by offering words of praise from its travelers. Forest Crossen rode the line in 1941, just before its abandonment:

Epilogue

> We crossed the dry arroyo into the outskirts of Santa Fe. Up winding Jefferson Street Engineer Albee took us, whistle blowing, bell clanging. And in the street stood children and women who waved greetings that touched my heart. Across San Francisco, the principal street running up to the historic Plaza, we crept … We crossed the trestle over the Santa Fe River and brushed by the weathered stone wall surrounding old Guadalupe Church with its graceful Lombardy Poplars … On and on across the narrow romantic streets- Agua Fria, De Vargas, Aztec and Montezuma Avenue—we moved, whistle and bell momentarily breaking the drowsy spell of Santa Fe, until we reached the station. I stepped down from the old chair coach of the 'Chili Line,' a glow of exultation going up over me. I was at once one with the gallant company of my countrymen who had ridden … over the narrow winding track of this pioneer railroad to Santa Fe.[4]

Then there is this from *The WPA Guide to 1930s New Mexico*:

> Approaching Santa Fe there is a view of the ancient city spread out below … Near the little station, the train crosses a trestle in front of the church of Our Lady of Guadalupe with brave sound of bell and show of steam and at the station the engineer stands proudly by while passengers descend.[5]

The generation that knew the Chili Line personally has perished. May their words remind us of this brave, proud little railroad, and those who worked so hard to develop and maintain it to the end.

APPENDIX

June 4, 1916 Chili Line Timetable[1]
Daily except Sunday

Southbound				Northbound
8:15	leave	Antonito	arrive	6:10 p.m
8:50		Palmilla		5:35
9:15		Volcano		5:10
9:45		No Agua		4:40
10:10		Tres Piedras		4:15
10:40		Servilleta		3:45
11:20		Taos Junction		3:10
11:45		Barranca		2:45
12:20	arrive	Embudo	leave	1:50
12:45	leave	Embudo	arrive	1;25
1:20		Alcade		12:45
1:45		Chamita		12:25
2:00		Española		12:10
2:20		San Ildefonso		11:50
2:35		Otowi		11:35
2:50		Buckman		11:20
3:30		Jacona		10:45
4:25	arrive	Santa Fe	leave	10:05

1. Adapted from Chappell, op.cit., p. 28.

ENDNOTES

ACKNOWLEDGMENTS

1. www.cumbrestoltec.org.
2. Haywood, P., 'Richard Dorman, 1922–2010: Santa Fe Architect Built Lasting Legacy,' *The New Mexican* (Santa Fe: April 8, 2010)

INTRODUCTION

1. Athearn, R., *The Denver and Rio Grande Western Railroad* (Lincoln: University of Nebraska Press, 1977), p. 15.
2. *Ibid*., p. 29.
3. *Ibid*., p. 31.
4. *Ibid*.
5. Wilkins, T. E., *Colorado Railroads Chronological Development* (Boulder, CO: Pruett Publishing Company, 1974), p. 19.
6. Myrick, D. F., *New Mexico's Railroads: A Historical Survey* (Albuquerque: University of New Mexico Press, 1990), p. 111.
7. *Ibid*., p.115.
8. Athearn, *op. cit*., p. 316.

CHAPTER 1

1. Butler, M., *Great Sand Dunes National Park* (Charleston, SC: Arcadia Publishing, 2013), p. 8.
2. Chappell, G., *To Santa Fe By Narrow Gauge: The D & RG's "Chili Line"* (Golden, CO: Colorado Railroad Museum, 1988), p. 6.
3. Pelton, A. R., *San Luis Valley Illustrated* (Monte Vista, CO: Adobe Village Press, 2003, reprint of 1891 edition), p. 114.
4. *Ibid*., p. 115.

5. Simmons, V. M., *The San Luis Valley: Land of the Six-Armed Cross* (Boulder, CO: Pruett Publishing Company, 1979), p. 91.
6. Gjevre, J. A., *Chili Line: The Narrow Rail Trail to Santa Fe, Third Revised Edition* (Moorhead, MN: Agassiz Publications, 2008), p. 97.
7. Simmons, *op. cit.*, p. 91.
8. Noel, T. J. (intro.), *The WPA Guide To 1930s Colorado* (Lawrence: University Press of Kansas, 1987), p. 400.
9. Richardson, R. W., *Chasing Trains* (Denver: Sundance Publications, 1995), p. 263.
10. *Ibid.*, pp. 264–266.
11. *Ibid.*, pp. 263–267.

CHAPTER 2

1. Chappell, *To Santa Fe By Narrow Gauge*, p. 7.
2. The information here is adapted from Wilson, Spencer and Vernon J. Glover, *The Cumbres & Toltec Scenic Railroad: The Historic Preservation Study* (Albuquerque: University of New Mexico Press, 2001), pp. 62–63.
3. Gjevre, *Chili Line*, p. 31.
4. *Ibid.*, p. 98.
5. Julyan, R., *The Mountains of New Mexico* (Albuquerque: University of New Mexico Press, 2006), p. 67.
6. Information on D&RG Chili Line structures in this book comes from an 1891 D&RG publication titled *Denver & Rio Grande Railroad Bridges, Buildings and Other Structures*, published in Gjevre, *op. cit.*, p. 128, and from www.cumbrestoltec.org.
7. Information on siding capacity in this book is from a D&RG chart published in Chappell, *op. cit.* p. 28.
8. Quoted in Chappell, *op. cit.*, p. 13.
9. Julyan, R., *The Place Names of New Mexico* (Albuquerque: University of New Mexico Press, 1998), p. 336.
10. Simmons, M. (foreword), *The WPA Guide To 1930s New Mexico* (Tucson: University of Arizona Press, 1989), p. 306
11. Gulliford, A., "Love Stories at Tres Piedras: Aldo Leopold's Cabin In New Mexico," The Durango Herald, September 7, 2016.
12. Quoted in Gjevre, *op. cit.*, p. 3.
13. Logan, J.R. "Tres Piedras Cabin Offers Window into Wilderness Advocate Leopold's Life," The Taos News, October 26, 2016.
14. Gulliford, *op. cit.*
15. *Ibid.*
16. www.aldoleopold.org
17. Quoted in Logan, *op. cit.*

CHAPTER 3

1. Sanders, G. E., *Oscar E. Berninghaus: Master Painter of American Indians and the Frontier West* (Taos, NM: Taos Heritage Publishing Company, 1985), p. 8.

2. *Ibid.*
3. *Ibid.*
4. Simmons, M., *The WPA Guide To 1930s New Mexico*, p. 412.
5. Gjevre, *Chili Line*, p. 79.
6. *Ibid.*, p. 108.
7. Myrick, *New Mexico's Railroads*, p. 122.
8. Gjevre, *op. cit.*, p. 80.
9. *Ibid.*
10. *Ibid.*, p. 108.
11. Myrick, *op. cit.*, p. 228.
12. *Ibid.*
13. Gjevre, *op. cit.*, p. 109.
14. Evans, M., *Long John Dunn of Taos* (Santa Fe, NM: Clear Light Publishers, 1993), p. 103.
15. *Ibid.*, p. 113.
16. Gjevre, *op. cit.*, p. 77.

CHAPTER 4

1. Gjevre, J. A., *Chili Line: The Narrow Rail Trail to Santa Fe*, First Edition, (Española, NM: Rio Grande Sun Press, 1969), p. 28.
2. Butler, M., *Taos: A Pictorial Guide for Travelers* (Santa Fe, NM: Sunstone Press, 2019), p. 24.
3. Chappell, *To Santa Fe By Narrow Gauge*, p. 7.
4. Julyan, *Place Names*, pp. 110–111.
5. Gjevre, *First Edition, op. cit.*, p. 32.
6. Gjevre, *Third Revised Edition, op. cit.*, p. 64.
7. Brown, J., "Westward Flow: The Embudo, New Mexico Stream-Gauging Station," *Civil Engineering*, July–August 2015, p. 19.
8. Frazier, A. H. and Wilbur H., *Embudo, New Mexico, Birthplace of Systematic Stream Gaging*, (Washington, D.C.: U.S. Government Printing Office, 1972), p. 5.
9. *Ibid.*
10. *Ibid.*, pp. 7–8.
11. Dutton, C. E., quoted in Frazier, *op. cit.*, p. 6.
12. Frazier, *op. cit.*, p. 20.

CHAPTER 5

1. Chappell, *To Santa Fe By Narrow Gauge*, p. 7.
2. *Ibid.*
3. Quoted in Gjevre, *Third Revised Edition*, p. 52.
4. Julyan, *Place Names*, p. 372.
5. *Ibid.*, p. 313.
6. *Ibid.*, p. 126.
7. Simmons, M., *The WPA Guide To 1930s New Mexico*, p. 414.
8. Quoted in Gjevre, *Third Revised Edition*, p. 53.

9. *Ibid.*, p. 50.
10. *Ibid.*, p. 22.
11. DeBuys, W. and Usner, D. J., *Valles Caldera: A Vision For New Mexico's National Preserve* (Santa Fe: Museum of New Mexico Press, 2006), p. 29.

CHAPTER 6

1. Gibson, D., *Pueblos of the Rio Grande: A Visitor's Guide* (Tucson, AZ: Rio Nuevo Publishers, 2001), p. 70.
2. Julyan, *Place Names*, p. 253.
3. LaFarge, O., *Santa Fe, The Autobiography of a Southwestern Town* (Norman: University of Oklahoma Press, 1939), pp. 356–358.
4. National Park Service, *National Register of Historic Places Inventory Nomination Form, Otowi, NM* (Washington: U.S. Department of the Interior, 1975), p. 2.
5. Cash, M. R., *Built of Earth and Song: Churches of Northern New Mexico* (Santa Fe, NM: Red Crane Books, 1993), pp. 34–35.
6. National Park Service, *op. cit.*, p. 6.
7. Oppenheimer, J. R., *Letter to Peggy Pond Church, November 21, 1958* (Albuquerque: University of New Mexico Libraries, Center for Southwest Research, Peggy Pond Church Papers). Box 1, Folder 9.
8. National Park Service, *op. cit.*, pp. 6–7.
9. Church, P. P., *The House at Otowi Bridge: The Story of Edith Warner and Los Alamos.* (Albuquerque: University of New Mexico Press, 1959), p. 4. This is an excellent biography of Edith Warner, and it interweaves Peggy Pond Church's life at Los Alamos Boys' Ranch School with Edith at Otowi. Peggy's father, Ashley Pond, established the school in 1917. In 1924, Peggy married Fermor Spencer Church, who became a teacher at the school, and later headmaster. The Church's stayed at Los Alamos until the school was closed in December 1942. During that time, they had many visits to Edith Warner's tearoom.

CHAPTER 7

1. Quoted in Gjevre, *Third Revised Edition*, p. 107.
2. *Ibid.*, p. 26.
3. Quoted in Steele, A. R., *Santa Fe 1880: Chronicles from the Year of the Railroad.* (Charleston, SC: The History Press, 2019), p. 11.
4. Myrick, *New Mexico's Railroads*, p. 7.
5. Historic Santa Fe Foundation, *Old Santa Fe Today, Fourth Edition.* (Albuquerque: University of New Mexico Press, 1991), p. 63.
6. Simmons, M., *The WPA Guide To 1930s New Mexico*, p. 194.
7. Dorman, R. L., *The Chili Line and Santa Fe the City Different* (Santa Fe, NM: R.D. Publications, Inc., 2000), p. 30.
8. www.sanmiguelchapel.org
9. Historic Santa Fe Foundation, *op. cit.*, p. 95.
10. Cash, *Built of Earth and Song*, p. 15.

EPILOGUE

1. Gjevre, *Third Revised Edition*, p. 17.
2. *Ibid.*, p. 116.
3. *Ibid.*
4. *Ibid.*, p. 107.
5. Simmons, M., *The WPA Guide To 1930s New Mexico*, p. 414.

BIBLIOGRAPHY

aldoleopold.org

Athearn, R., *The Denver and Rio Grande Western Railroad* (Lincoln: University of Nebraska Press, 1977)

Brown, J., 'Westward Flow: The Embudo, New Mexico Stream Gauging Station,' *Civil Engineering* (July–August 2015)

Butler, M., *Great Sand Dunes National Park* (Charleston, SC: Arcadia Publishing, 2013); *Taos: A Pictorial Guide for Travelers* (Santa Fe, NM: Sunstone Press, 2019)

Cash, M. R., *Built of Earth and Song: Churches of Northern New Mexico* (Santa Fe, NM: Red Crane Books, 1993)

Chappell, G., *To Santa Fe By Narrow Gauge: The D & RG's "Chili Line"* (Golden, CO: Colorado Railroad Museum, 1988)

Church, P. P., *The House at Otowi Bridge: The Story of Edith Warner and Los Alamos* (Albuquerque: University of New Mexico Press, 1959)

cumbrestoltec.org

DeBuys, W. and Usner, D. J., *Valles Caldera: A Vision For New Mexico's National Preserve* (Santa Fe: Museum of New Mexico Press, 2006)

Dorman, R. L., *The Chili Line and Santa Fe the City Different* (Santa Fe, NM: R.D. Publications, Inc., 2000)

Evans, M., *Long John Dunn of Taos* (Santa Fe, NM: Clear Light Publishers, 1993)

Frazier, A. H. and Heckler, W., *Embudo, New Mexico, Birthplace of Systematic Stream Gaging* (Washington: U.S. Government Printing Office, 1972)

Gibson, D. P., *Pueblos of the Rio Grande: A Visitor's Guide* (Tucson, AZ: Rio Nuevo Publishers, 2001)

Gjevre, J. A., *Chili Line: The Narrow Rail Trail to Santa Fe, First Edition* (Española, NM: Rio Grande Sun Press, 1969); *Chili Line: The Narrow Rail Trail to Santa Fe, Third Revised Edition* (Moorhead, MN: Agassiz Publications, 2008).

Gulliford, A., 'Love Stories At Tres Piedras: Aldo Leopold's Cabin In New Mexico,' *The Durango Herald*, (Durango, CO: September 7, 2016)

Haywood, P., 'Richard Dorman, 1922-2010: Santa Fe Architect Built Lasting Legacy,' *The New Mexican* (Santa Fe, April 8, 2010)

Historic Santa Fe Foundation, *Old Santa Fe Today, Fourth Edition* (Albuquerque: University of New Mexico Press, 1991)

Julyan, R., *The Mountains of New Mexico* (Albuquerque: University of New Mexico Press, 2006); *The Place Names of New Mexico* (Albuquerque: University of New Mexico Press, 1998)

LaFarge, O., *Santa Fe, The Autobiography of a Southwestern Town* (Norman: University of Oklahoma Press, 1939)

Logan, J. R., 'Tres Piedras Cabin Offers Window Into Wilderness Advocate Leopold's Life,' *The Taos News*, (Taos, NM: October 26, 2016)

Myrick, D. F., *New Mexico's Railroads: A Historical Survey* (Albuquerque: University of New Mexico Press, 1990)

National Park Service, 'National Register of Historic Places Inventory Nomination Form, Otowi, NM,' *U.S. Department of the Interior* (Washington: 1975)

Noel, T. J. (introduction), *The WPA Guide to 1930s Colorado* (Lawrence: University Press of Kansas, 1987)

Oppenheimer, J. R., 'Letter To Peggy Pond Church, November 21, 1958,' *University of New Mexico Libraries, Center for Southwest Research*, (Albuquerque: 1958)

Pelton, A. R., *San Luis Valley Illustrated, Reprint of 1891 Edition* (Monte Vista, CO: Adobe Village Press, 2003)

Richardson, R. W., *Chasing Trains* (Denver, CO: Sundance Publications, 1995)

Sanders, G. E., *Oscar E. Berninghaus: Master Painter of American Indians and the Frontier West* (Taos, NM: Taos Heritage Publishing Company, 1985)

sanmiguelchapel.org

Simmons, M. (foreword), *The WPA Guide to 1930s New Mexico* (Tucson: University of Arizona Press, 1989)

Simmons, V. M., *The San Luis Valley: Land of the Six-Armed Cross* (Boulder, CO: Pruett Publishing Company, 1979)

Steele, A. R., *Santa Fe 1880: Chronicles from the Year of the Railroad* (Charleston, SC: The History Press, 2019)

Wilkins, T. E. *Colorado Railroads Chronological Development* (Boulder, CO: Pruett Publishing Company, 1974)

Wilson, S. and Glover, V. J., *The Cumbres and Toltec Scenic Railroad: The Historic Preservation Study* (Albuquerque: University of New Mexico Press, 2001)